A Beginner's Guide

to

Magic Squares & Cubes

By

Kenneth Kelsey

K J Kelsey
kelsey@talk21.com

Cover design by
Jo Kelsey
jo1208@hotmail.com

Contents

MELENCOLIA by Albrecht Dürer, 1514

Foreword

In the 1970s I had occasion to call upon one of the Directors of the Rijksgebouwendienst in The Hague. On the wall of his office was a carved stone plaque, retrieved from a derelict seventeenth century Dutch farmhouse. It consisted of the following square,

16	3	2	13
5	10	11	8
9	6	7	12
4	15	14	1

and included the carved word 'Compatibility'.

The Director explained that the square's compatibility arose from the fact that the figures 1 to 16 were so arranged within the square that the rows, columns and diagonals all added up to 34.

Surprisingly, this was my first introduction to a Magic Square, since they have been around for two or three thousand years. An identical plaque is depicted in a 1514 engraving by the German Renaissance master, Albrecht Dürer, in the Staatsgalerie in Stuttgart, which depicts the Magic Square near the top right hand corner. However, the message Dürer was sending was not 'Compatibility' but 'Melancholy', for in the left hand corner he had engraved the word 'MELENCOLIA'.

*

So what's this book about? Well, the clue is in the title. It's a beginner's guide to a numerical peculiarity, and since it's for beginners there will be no formulas, no algebra, no algorithms and no computer programming within its pages. All you will need in order to enjoy and understand it is some simple arithmetic, a modicum of logic and a growing sense of wonder.

What are Magic Squares? A regular Magic Square comprises a square composed of consecutive numbers arranged in such a way that the rows, columns and diagonals add up to the same total or constant. The smallest such square is 3x3. Beyond that there is no limit to their size, but the largest we will be concerned with in this book is 9x9.

What are Magic Cubes? They consist of a series of Magic Squares which, taken as a whole, are composed of consecutive numbers and whose rows, columns and diagonals add up to the same constant not only in each planar square but also through the cube.

Some Magic Squares and Cubes possess additional characteristics which enable the constant to be obtained from several other groupings and which warrant the Square or Cube being re-designated as near-perfect or perfect. These will be elaborated upon in the following chapters.

Do Magic Squares and Cubes have a use? From a practical point of view the answer is 'no'. They are merely fun. They engage the mind, as do crosswords and Sudoku, in solving the many puzzles that can be set using them, and the satisfaction obtained in devising them, for Magic Squares do not necessarily have to be square, as we shall see. As you proceed through the Chapters I believe that you will sense the harmony, the balance, the logic and the stability of numbers, and their universal and enduring application.

Dürer's Magic Square is as valid today as it was 500 years ago and will be 500 years hence.

* * *

BOOK ONE

MAGIC SQUARES

Preamble. Much has been written about Magic Squares and their properties, what defines a particular magic square and how its characteristics should be classified. It is not my intention to confuse matters further but to simplify things for the beginner. In this book I will categorise only three types of Magic Square:

Basic: where the constant is produced only by its rows, columns and two diagonals.

Near perfect: where the constant is produced by additional structural features.

Perfect: where no further structural features are possible to produce the constant.

*

Subsets are non-linear groupings of cells which produce the constant as a direct result of the construction of the Magic Square, not fortuitously.

As the constant is the common feature of all Magic Squares I will use the number of ways a square produces the constant as the means of comparison between squares. That detail will be shown in 'Constant Count' boxes after each square in this book as appropriate.

* * *

3x3 Magic Squares

This is the smallest possible magic square and contains the numbers 1 to 9 arranged in such a way that they produce a constant of 15.

2	9	4	15
7	5	3	15
6	1	8	15
15	15	15	

It is of ancient origin. The Chinese call it Lo Shu.

There is more than one story of its origin. One has it that the Emperor saw it on the carapace of a sacred turtle as he walked along the banks of the river Lo. Hence Lo Shu, the diagram of the river Lo.

Another version has it that once there was a huge flood. The people tried to offer some sacrifice to the god of one of the flooding rivers, the river Lo, to calm his anger. However, each time they did so a turtle came from the river and walked around the sacrifice, a sign that the river god did not accept the sacrifice. Eventually a child noticed the curious Lo Shu symbol on the turtle's back, and the people realized that the correct amount of sacrifice needed was 15.

The 3x3 Magic Square made its way out of China and entered the Indian subcontinent. From India, it travelled on to Arabia and into medieval Europe. Since then, Magic Squares have fascinated humanity throughout the ages. They are found in a number of cultures, including Egypt and India, engraved on stone or metal and worn as talismans, the belief being that Magic Squares had astrological and divinatory qualities, their usage ensuring longevity and prevention of diseases. In the ninth century, Arabian astrologers used Magic Squares to interpret horoscopes.

Apart from displaying the Lo Shu square in 8 forms by rotations and reflections, little more can be said about it at this stage, though its significance will be recognised as crucial when we come to 9x9 pandiagonal squares later in the book.

* * *

4x4 Magic Squares

These are squares in which the numbers 1 to 16 are so arranged that the rows, columns and diagonals all produce the constant of 34.

There are many ways in which Magic Squares can be created. The simplest way of creating a 4x4 square is to arrange the figures in arithmetical sequence and then to reverse the diagonals.

1	2	3	4
5	6	7	8
9	10	11	12
13	14	15	16

16	2	3	13
5	11	10	8
9	7	6	12
4	14	15	1

This produces a near perfect Magic Square with a constant of 34. You will no doubt recognise it as similar to Dürer's square, where he has switched columns 2 and 3 in order to have the date, 1514, read in the bottom row, a move which surprisingly still left the diagonal constants untouched.

16	3	2	13
5	10	11	8
9	6	7	12
4	15	14	1

Durer's square

If he had realised that surprising fact and pondered upon its implications, would that have mellowed his melancholia? We will never know.

Another way to form a Magic Square involves using Latin squares – squares in which each number or letter appears only once in each row or column. A very common form is found in Sudoku puzzles where, in the 9x9 grid, each number appears once, not only in each row or column but also in each 3x3 grid. The method of using Latin squares to create a Magic Square is as follows:

Here are two Latin squares with letters in place of numbers:

A	B	C	D
D	C	B	A
B	A	D	C
C	D	A	B

w	z	x	y
x	y	w	z
y	x	z	w
z	w	y	x

Let us now place numerical values to those letters:

A = 1	w = 0
B = 2	x = 4
C = 3	y = 8
D = 4	z = 12

Adding the two squares together produces:

1	14	7	12	34
8	11	2	13	34
10	5	16	3	34
15	4	9	6	34
34	34	34	34	

Fig 1

All rows, columns and diagonals in this square produce the constant of 34, but there is far more to the square than that. In creating the two Latin squares above I made sure that each letter appeared in the diagonals only once, which has the effect of the completed Magic Square being pandiagonal. **Pandiagonal** squares possess additional qualities: The broken diagonals, e.g. 14, 2, 3 and 15, all produce the constant. In addition one or more rows can be moved from the bottom to the top and/or one or more columns can be moved from the right to the left without destroying the diagonal constants. If we examine this Magic Square further we can see that any subset of four cells which together form a square, whether within the square's perimeter or beyond it, (e.g. 4, 9, 14 and 7) also produce the constant.

15	4	9	6	15
1	14	7	12	1
8	11	2	13	8
10	5	16	3	10
15	4	9	6	15

Fig 2

There are no other possible groupings of 34 so the above square rates as a perfect Magic Square.

Constant Count

Rows	4
Columns	4
Fwd Diagonals	4
Bwd Diagonals	4
Subsets	16
Total	32

How many 4x4 Magic Squares are there? In this book I am concerned only with Magic Squares created with consecutive numbers or series and which are created in a uniform pattern. The class into which a Magic Square falls is determined by the relative position of the reciprocals, i.e. 1 and 16; 2 and 15 etc in a 4x4 square. There are 9 such classes, detailed below, Class B being the pandiagonal form.

10

Square A

1	14	15	4
12	7	6	9
8	11	10	5
13	2	3	16

A

Square B

1	14	11	8
12	7	2	13
6	9	16	3
15	4	5	10

B

Square C

1	11	8	14
6	16	3	9
12	2	10	7
15	5	10	4

C

Square D

1	11	6	16
8	14	3	9
12	2	15	5
13	7	10	4

D

Square E

1	14	15	4
7	12	9	6
10	5	8	11
16	3	2	13

E

Square F

1	11	16	6
8	14	9	3
10	4	7	13
15	5	2	12

F

Square G

1	14	11	8
4	15	10	5
16	3	6	9
13	2	7	12

G

Square H

1	16	11	6
4	13	10	7
14	3	8	9
15	2	5	12

H

Square I

1	11	8	14
16	6	9	3
10	4	15	5
7	13	2	12

I

Fig 3

Let us examine the Class A example, highlighted below. If we retain 1 and 16 in their current locations, we can re-position the other numbers in 24 ways while still retaining the Class characteristics, so with the number 1 in the first cell and 16 in the last, there are 24 possible Magic Squares, as shown in the following diagram:

1	1	1	1	1	1	1	1	1	1	1	1	1	1	1	1	1	1	1	1	1	1	1	1
8	8	8	8	8	8	12	12	12	12	12	12	14	14	14	14	14	14	15	15	15	15	15	15
12	12	14	14	15	15	8	8	14	14	15	15	8	8	12	12	15	15	8	8	12	12	14	14
13	13	11	11	10	10	13	13	7	7	6	6	11	11	7	7	4	4	10	10	6	6	4	4
15	14	12	15	12	14	14	15	8	15	8	14	12	15	8	15	12	8	12	14	8	14	12	8
10	11	13	10	13	11	7	6	13	6	13	7	7	4	11	4	7	11	6	4	10	4	6	10
6	7	7	4	6	4	11	10	11	4	10	4	13	10	13	6	6	10	13	11	13	7	7	11
3	2	2	5	3	5	2	3	2	9	3	9	2	5	2	9	9	5	3	5	3	9	9	5
14	15	15	12	14	12	15	14	15	8	14	8	15	12	15	8	8	12	14	12	14	8	8	12
11	10	10	13	11	13	6	7	6	13	7	13	4	7	4	11	11	7	4	6	4	10	10	6
7	6	4	7	4	6	10	11	4	11	4	10	10	13	6	13	10	6	11	13	7	13	11	7
2	3	5	2	5	3	3	2	9	2	9	3	5	2	9	2	5	9	5	3	9	3	5	9
4	4	6	6	7	7	4	4	10	10	11	11	6	6	10	10	13	13	7	7	11	11	13	13
5	5	3	3	2	2	9	9	3	3	2	2	9	9	5	5	2	2	9	9	5	5	3	3
9	9	9	9	9	9	5	5	5	5	5	5	3	3	3	3	3	3	2	2	2	2	2	2
16	16	16	16	16	16	16	16	16	16	16	16	16	16	16	16	16	16	16	16	16	16	16	16

136 136 136 136 136 136 136 136 136 136 136 136 136 136 136 136 136 136 136 136 136 136 136 136

Further, each of the other 15 numbers can occupy the first cell in turn, so the possibilities increase to 24 x 16, which totals 384 for each Class. There are 9 Classes, so the total possible number of 4x4 Magic Squares is 384 x 9 = 3,456, a figure which includes rotations and mirror images.

I need to comment upon the obvious similarities between some of the classes. By rotation and/or reflection Class D can be converted to Class E; Class F to Class G and Class H to Class I. That does not mean that this thereby reduces the number of classes to six, for that would imply that there are only six cells for a reciprocal to occupy. The diagram above shows that there are nine – and only nine - an aspect which intrigues me. In all nine Classes the reciprocals are in a linear relationship. This raises the question, is it mathematically impossible for them to be related by, for example, a Knight's move? I tried to create a Magic Square by the Latin Square method, with 1 in cell one and 16 in cell seven but found it impossible. Every attempt produced duplicate numbers. This leads me to believe that reciprocals in even-numbered Magic Squares must always be in a linear relationship. We will return to linear relationships later in this book but in a different context.

* * *

Puzzle Corner

A Caliph, renowned for his cunning, once said to his tailor, "You say I owe you more, yet I say I owe you but 34 gold pieces. Since we cannot agree and I am a generous man, I will allow you to make a greater profit if Allah so wills. Here are 16 boxes, arranged in four rows of four. See, I am placing 1 gold coin in this box, 2 in this box, 3 in this box, and so on to 16 in this last box. You can see where all the gold coins lie. You may choose any four boxes in a row, be it horizontal, vertical or diagonal."

The tailor studied the boxes and after a while said, "Sire, it matters not which row I choose, for they all contain 34 gold pieces."

"My generosity is boundless," said the Caliph. "You may choose any four boxes which together form a square."

The tailor studied the boxes again, then said, "Sire, all such boxes contain exactly 34 gold coins."

"I am just, as well as generous," said the Caliph. "Before you choose you may move any number of rows from the left side to the right, or from the top to the bottom, or do both."

The tailor studied the boxes yet again. "Sire," he said at last, "your generosity I question not, yet no matter how I move the boxes I cannot change the number of gold coins in any such grouping of four. Truly has Allah shown me that you owe me but 34 pieces of gold."

In which order had the cunning Caliph placed the coins in the boxes? The contents of five boxes are shown.

1			
		3	
			2
14			11

Puzzle No. 1

5x5 Magic Squares

These are squares in which the numbers 1 to 25 are so arranged that the rows, columns and diagonals all produce the constant of 65.   As was said earlier, there are many ways to create Magic Squares, and the method outlined below can be applied to most odd-numbered squares. First, place the 25 figures in a grid as outlined,

			5			
		4		10		
	3		9		15	
2		8		14		20
1	7		13		19	25
6		12		18		24
	11		17		23	
		16		22		
			21			

Fig 4

then transfer the figures outside the centre grid to the furthest vacant square, horizontally or vertically as the case may be.

3	16	9	22	15	65
20	8	21	14	2	65
7	25	13	1	19	65
24	12	5	18	6	65
11	4	17	10	23	65
65	65	65	65	65	

Fig 5

This produces a basic Magic Square.

Converting it into a pandiagonal Magic Square can be achieved by transposing the row order to 1, 3, 5, 2, 4, thus:

3	16	9	22	15	65
7	25	13	1	19	65
11	4	17	10	23	65
20	8	21	14	2	65
24	12	5	18	6	65
65	65	65	65	65	

Fig 6

We now have a pandiagonal square in which the following groupings all produce a constant of 65. There are also subsets which produce the constant, namely any five numbers in cross formation, either upright or slanting.

There are no other possible subsets which could produce the constant, so the above square rates as a perfect Magic Square.

Constant Count

Rows	5
Columns	5
Fwd Diagonals	5
Bwd Diagonals	5
Subsets	50
	___
Total	70
	===

* * *

Puzzle Corner

The form-master was addressing his class.

"The Head is arranging a tug-of-war competition between the three classes in this year, and to avoid any class putting up their five heaviest pupils he himself is going to select each team at random by choosing five boys in a row, either horizontally, vertically or diagonally across the classroom. Now there is no point in my putting the five heaviest boys in one particular row, hoping the Head will choose that one, because if he doesn't we will have lessened our chances. So what I have decided to do is this. I will grade you all by weight, the heaviest being number 1 and so on down to the lightest at number 25. Then I will position you throughout the classroom so that whichever row the Head selects, he will be choosing a perfectly average weight of boys."

How did the form-master arrange the class? The position and grading of eight of the boys are given.

	20		22	
	25		21	
				17
24				
14				18

Puzzle No. 2

6x6 Magic Squares

Squares of this order, and indeed orders of the 10th, 14th and 18th and so on, are difficult to construct and have no pandiagonal versions with consecutive numbers. However, it is possible to construct a 6x6 Pandiagonal Magic Square with a broken series, as is shown below, but it is highly unlikely that this could be extrapolated to a successful Magic Cube.

35	29	42	1	7	36	150
16	20	9	48	44	13	150
31	33	38	5	3	40	150
17	19	10	47	45	12	150
30	34	37	6	2	41	150
21	15	14	43	49	8	150
150	150	150	150	150	150	

Fig 7

Another construction problem is that the Latin Square method is unavailable for 6 power Magic Squares. At first glance it might seem possible to take a 4th order square and construct a numerical surround for it, but one look at the arithmetic reveals that this is starkly impossible. However, the idea of a surround need not be discarded.

Take a 4th order square

1	15	10	8	34
12	6	3	13	34
7	9	16	2	34
14	4	5	11	34
34	34	34	34	

Fig 8

and increase all the values by 10:

	11	25	20	18	
	22	16	13	23	
	17	19	26	12	
	24	14	15	21	

Fig 9

We now have twenty numbers available with which to create the surround - 1 to 10 and 27 to 36. From those 20 numbers we have to select 10 pairs which total 37, for each row, column and diagonal. That's the easy part. Those pairs have to be so arranged that they total 111 in rows 1 and 6, columns 1 and 6 and the 2 diagonals. There are nearly 500 such possibilities, so with a little effort we can produce the following, but the square will not be pandiagonal:

35	28	3	4	5	36	111
6	11	25	20	18	31	111
10	22	16	13	23	27	111
30	17	19	26	12	7	111
29	24	14	15	21	8	111
1	9	34	33	32	2	111
111	111	111	111	111	111	

Fig 10

There is little more to be said about 6x6 Magic Squares, so on to the puzzle.

Puzzle Corner

As I am unable to set a pandiagonal puzzle of the 6th order I will tell you a cautionary tale which you will not find in any history book.

A professor in charge of an archaeological dig gently brushed the soil from the tablet he had just uncovered.

"This is most interesting," he said. "It records that the Emperor was to pass in a straight line through the city on his 38th birthday, and to honour the occasion the Provincial Governor, Numericus Fastidius, placed a different number of the Legion's 190 standards at each road intersection so that whichever route the Emperor chose for his imperial progress through the city, he would pass exactly 38 standards. Then there is a diagram of the city showing the 19 road intersections and the number of standards at each of them, though unfortunately only 6 of them are clearly decipherable." He brushed some more soil away.

"Oh dear!" he said. "It then goes on to say that the Emperor was so flattered by the compliment that he returned on his following birthday, and when Numericus failed to repeat the tribute he had him thrown to the lions!"

The city diagram showing the 6 decipherable numbers is depicted below. Can you, without fear of furnishing a feline feast, provide the others?

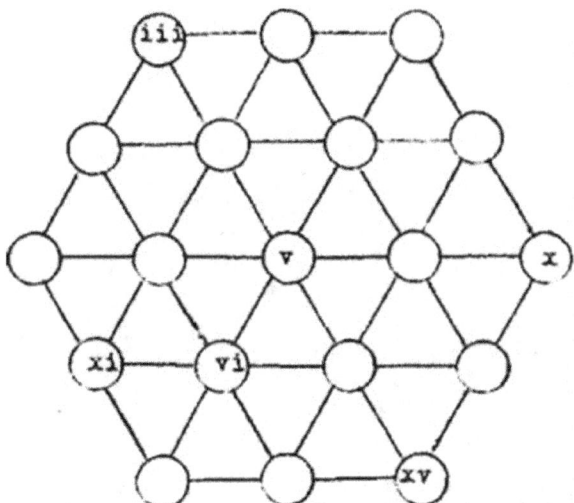

Only kidding! Sorry!

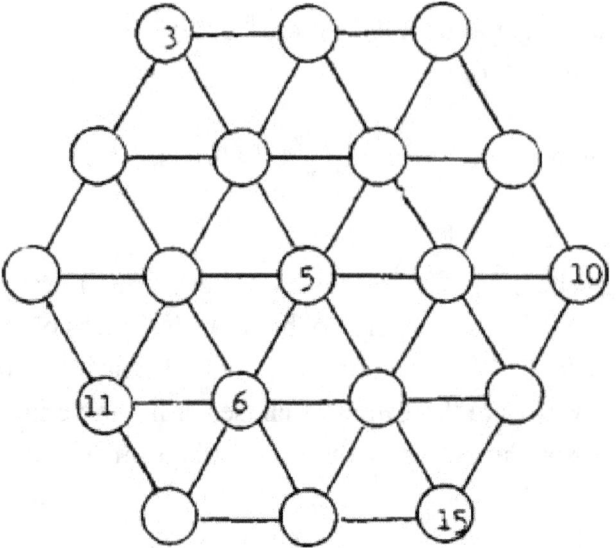

Puzzle No. 3

Poor old Numericus found to his cost that this cross-sum configuration is unique to the number 38.

7x7 Magic Squares

I wonder why seven is such a favoured number. Sailors sail the seven seas, there are seven deadly sins, Seven Wonders of the Ancient World, seven colours in the rainbow, seven days in a week, seven samurai, the Seven Sisters on the South Coast, a seventh heaven and, of course, seven brides for seven brothers. Is it a lucky number? We shall see.

7x7 Magic Squares can be created by the method outlined in the Chapter on 5x5 Magic Squares:

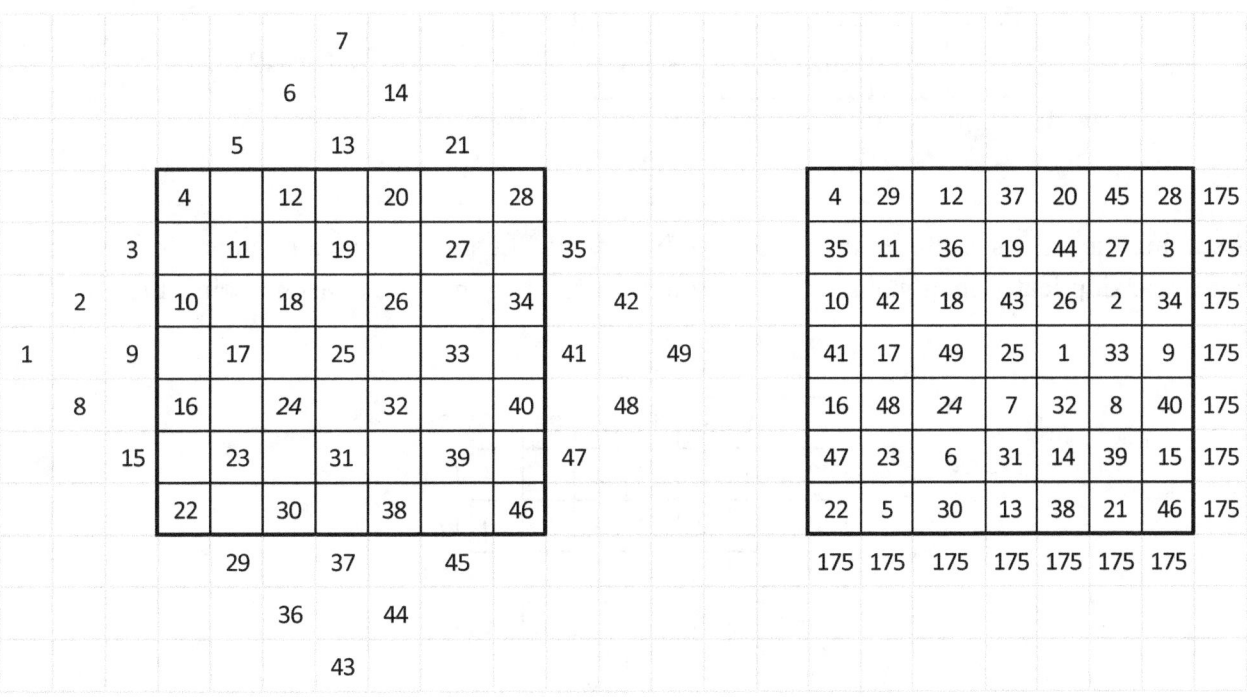

Fig 11

This has not produced a pandiagonal square, so using the Latin Square method let's construct another using the following grid:

22

At	Dv	Gx	Cz	Fu	Bw	Ey
Cw	Fy	Bt	Ev	Ax	Dz	Gu
Ez	Au	Dw	Gy	Ct	Fv	Bx
Gv	Cx	Fz	Bu	Ew	Ay	Dt
By	*Et*	Av	Dx	Gz	Cu	Fw
Du	Gw	Cy	Ft	Bv	Ex	Az
Fx	Bz	Eu	Aw	Dy	Gt	Cv

A=1   t=0
B=2   u=7
C=3   v=14
D=4   w=21
E=5   *x=28*
F=6   y=35
G=7   z=42

Fig 12

Applying the values produces the following 7x7 Magic Square, and since no upper or lower case letter was duplicated in any diagonal, this square is pandiagonal, the constant being 175.

1	18	35	45	13	23	40	175
24	41	2	19	29	46	14	175
47	8	25	42	3	20	30	175
21	31	48	9	26	36	4	175
37	5	15	32	49	10	27	175
11	28	38	6	16	33	43	175
34	44	12	22	39	7	17	175
175	175	175	175	175	175	175	

Fig 13

We now have a pandiagonal square in which the following groupings all produce a constant of 175. The subsets are numbers in 3-1-3 formation either horizontally or vertically.

```
┌────────────────────────────────────────┐
│           Constant Count                │
│                                         │
│   Rows                    7             │
│   Columns                 7             │
│   Fwd Diagonals           7             │
│   Bwd Diagonals           7             │
│   Subsets                84             │
│                        _____            │
│   Total                 112             │
│                        =====            │
│                                         │
└────────────────────────────────────────┘
```

The above square clearly rates as a perfect Magic Square.

* * *

Puzzle Corner

My Aunt Agatha, a noted numerologist in her day, once told me an amusing story concerning some workmen she had employed to re-decorate her study. It appears that she had a favourite Magic Square in which the numbers 1 to 16 were so arranged that all the horizontal, vertical and diagonal lines totalled 34, and she wished this Magic Square and its three rotations through 90, 180 and 270 degrees to be depicted in her study, one on each of the four walls. She explained this to the workmen and gave them a diagram of each of the four squares. When they had gone to lunch, she looked in to examine their handiwork and found that their painting had reached the stage depicted below, with only two or three numbers having been painted in on three of the squares, the remaining wall being still completely blank. She picked up the diagrams to check the accuracy of their painting but absent-mindedly put them into her handbag instead of replacing them on the workbench. It was while she was out shopping much later in the afternoon that she discovered that she still had the working diagrams and immediately took a taxi home so that the decorators could continue with their work. Rushing breathlessly into the study she was astonished to find that, despite the absence of the diagrams, the workmen had successfully reconstructed all four squares and had completed their painting. From the information given can you reproduce the Magic Square?

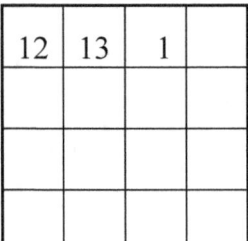 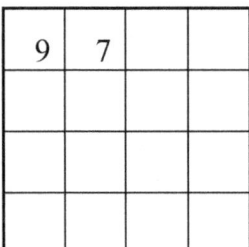 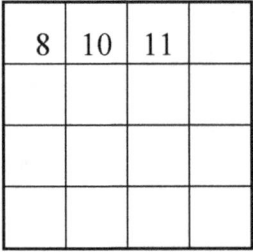

Puzzle No. 4

8x8 Magic Squares

The simplest way of creating a regular Magic Square of the 8th order is to insert the numbers in sequence in an 8x8 grid and reverse the diagonals in blocks of four.

1	2	3	4	5	6	7	8
9	10	11	12	13	14	15	16
17	18	19	20	21	22	23	24
25	26	27	28	29	30	31	32
33	34	35	36	37	38	39	40
41	42	43	44	45	46	47	48
49	50	51	52	53	54	55	56
57	58	59	60	61	62	63	64

64	63	3	4	5	6	58	57
56	55	11	12	13	14	50	49
17	18	46	45	44	43	23	24
25	26	38	37	36	35	31	32
33	34	30	29	28	27	39	40
41	42	22	21	20	19	47	48
16	15	51	52	53	54	10	9
8	7	59	60	61	62	2	1

Fig 14

The constant of the resulting square is 260 but the square is not pandiagonal. We can always turn to the Latin square method to produce a pandiagonal square. Here are the two primary squares I have chosen:

A	F	G	D	C	H	E	B
G	D	A	F	E	B	C	H
B	E	H	C	D	G	F	A
H	C	B	E	F	A	D	G
B	E	H	C	D	G	F	A
H	C	B	E	F	A	D	G
A	F	G	D	C	H	E	B
G	D	A	F	E	B	C	H

s	t	y	z	w	x	u	v
t	s	z	y	x	w	v	u
z	y	t	s	v	u	x	w
y	z	s	t	u	v	w	x
u	v	w	x	y	z	s	t
v	u	x	w	z	y	t	s
x	w	v	u	t	s	z	y
w	x	u	v	s	t	y	z

A=1	s=0
B=2	t=8
C=3	u=16
D=4	v=24
E=5	w=32
F=6	x=40
G=7	y=48
H=8	z=56

Fig 15

Applying the values produces the following 8th order pandiagonal square

1	14	55	60	35	48	21	26	260
15	4	57	54	45	34	27	24	260
58	53	16	3	28	23	46	33	260
56	59	2	13	22	25	36	47	260
18	29	40	43	52	63	6	9	260
32	19	42	37	62	49	12	7	260
41	38	31	20	11	8	61	50	260
39	44	17	30	5	10	51	64	260
260	260	260	260	260	260	260	260	

Fig 16

in which the following groupings produce a constant of 260.

Constant Count

Rows	8
Columns	8
Fwd Diagonals	8
Bwd Diagonals	8
Subsets of 2x4 numbers by row	64
Subsets of 2x4 numbers by column	64
Subsets of half row plus half row	104
Subsets of half column plus half column	104
Total	368

There are no further subsets possible, so the above square rates as a perfect Magic Square.

* * *

Puzzle Corner

"My liege," said the young knight, "set me a task, I pray you, that I might find honour, fame and fortune in your service."

The King therefore led him to a square courtyard paved as is a chequer-board, with tiles of black and white, sixty-four in number. The yard was bare, but in one corner stood a cell with door both barred and bolted.

The King pointed. "Sir Knight," he said, "know you that two years back my youngest daughter was imprisoned in yon cell, cast there under a spell by wicked Bandar. None can lift the spell and draw the bolts save any knight who, entering upon the courtyard here, and moving only as a chess knight moves, smites with his sword every tile in turn, the first tile once, the second twice, the third thrice and thus unto the sixty-fourth tile outside the cell. The sum of the number of smites in every row, both North to South and East to West must equal 260. Then, and only then will the spell be lifted and the bolts be drawn. Lift the spell and you shall have my daughter's hand in marriage, with honour, fame and fortune enough. Fail, and you will surely perish under Bandar's spell, as have the many brave knights who have gone before."

The young knight drew his sword, and stepping into the courtyard, smartly smote the first tile once …..

No royal hand in marriage awaits you if you succeed in tracing the young knight's steps to the cell door, but then, neither will you surely perish if you don't. To help you, though, the fourth of all his moves is given.

28

		49				17	64
1				33			
	29		45		61		13
		53		37			
			5		21		
		25		9			
			41		57		

Puzzle No. 5

9x9 Magic Squares

Nine is an extraordinary number. If a number, arithmetically known as the dividend, is exactly divisible by 9, then the digits of that dividend will successively add up to 9, as will those of the answer, the quotient. If the dividend cannot be divided exactly by 9, the quotient will end with a recurring decimal. If the digits of that dividend add up to 1, then the recurring decimal will be 1. If 2, then 2 – and so on. There are many more weird aspects of the number nine, but we will press on.

When I first began constructing number puzzles it was widely assumed that there was no firm method of creating 9th order pandiagonal magic squares containing the numbers 1 to 81, so I contented myself with making puzzles by manipulating nine 3x3 squares, such as the following, where I added 9 successively to the Lo Shu square and repositioned the resulting squares into Lo Shu order, producing a 9x9 Magic Square with a constant of 369.

2	9	4	11	18	13	20	27	22	126
7	5	3	16	14	12	25	23	21	126
6	1	8	15	10	17	24	19	26	126
29	36	31	38	45	40	47	54	49	369
34	32	30	43	41	39	52	50	48	369
33	28	35	42	37	44	51	46	53	369
56	63	58	65	72	67	74	81	76	612
61	59	57	70	68	66	79	77	75	612
60	55	62	69	64	71	78	73	80	612
288	288	288	369	369	369	450	450	450	

11	18	13	74	81	76	29	36	31	369
16	14	12	79	77	75	34	32	30	369
15	10	17	78	73	80	33	28	35	369
56	63	58	38	45	40	20	27	22	369
61	59	57	43	41	39	25	23	21	369
60	55	62	42	37	44	24	19	26	369
47	54	49	2	9	4	65	72	67	369
52	50	48	7	5	3	70	68	66	369
51	46	53	6	1	8	69	64	71	369
369	369	369	369	369	369	369	369	369	

Fig 17

Another method produced the following:

65	2	47	72	9	54	67	4	49	369
20	38	56	27	45	63	22	40	58	369
29	74	11	36	81	18	31	76	13	369
70	7	52	68	5	50	66	3	48	369
25	43	61	23	41	59	21	39	57	369
34	79	16	32	77	14	30	75	12	369
69	6	51	64	1	46	71	8	53	369
24	42	60	19	37	55	26	44	62	369
33	78	15	28	73	10	35	80	17	369
369	369	369	369	369	369	369	369	369	

Fig 18

You will no doubt detect the Lo Shu pattern within this square.

However, whilst I was in the Netherlands I came across a book entitled 'Domino Games and Domino Puzzles' by K.W.H. Leeflang, published in 1972. In it he set out very clearly the means of producing a 9x9 pandiagonal square which involved, as a first step, changing the row and column order, thus:

1	2	3	4	5	6	7	8	9		1
10	11	12	13	14	15	16	17	18		2
19	20	21	22	23	24	25	26	27		3
28	29	30	31	32	33	34	35	36		6
37	38	39	40	41	42	43	44	45		9
46	47	48	49	50	51	52	53	54		7
55	56	57	58	59	60	61	62	63		8
64	65	66	67	68	69	70	71	72		4
73	74	74	76	77	78	79	80	81		5

1	2	3	8	9	5	6	4	7
10	11	12	17	18	14	15	13	16
19	20	21	26	27	23	24	22	25
46	4	48	53	54	50	51	49	52
73	74	75	80	81	77	78	76	79
55	56	57	62	63	59	60	58	61
64	65	66	71	72	68	69	67	70
28	29	30	35	36	32	33	31	34
37	38	39	44	45	41	42	40	4

Fig 19

You will notice that he has chosen two sequence changers – 123 697 845 for the rows and 123 895 647 for the columns. The interesting thing about these two series of figures is that in each of them the first, fourth and seventh; the second, fifth and eighth; and the third, sixth and ninth all total 15. In other words the rows comprise three set of triplets (168, 294, 375) and the columns similarly comprise the triplets (186, 294, 357), the same figures but not in the same order. The numbers 1, 2 and 3 can never be paired in a triplet. In a footnote K.W.H. Leeflang said that there were 24 such sequence changers and since they are interchangeable it is possible to generate 24^2 such grids.

The grid, however, is not a magic square. To achieve this, the figures need to be redistributed in an orderly fashion in order to lead to the creation of a 9x9 pandiagonal Magic Square. Leeflang took the following route, though this was not the only route he could have chosen. His starting point for cell 1 was figure 75 (row 9 column 3). It did not matter which starting point he chose since the resulting pandiagonal square would allow any figure to occupy the first cell. He then chose the next 8 figures of row 1 by successively moving down 2 rows and 6 further cells to the right (d2/6r). Having reached the end of row 1 he then chose a fresh starting point for row 2 by moving up one row and 5 further cells to the left, (u1/5l) which turned out to be 10. He then returned to the sequence, (d2/6r). Having reached the end of row 2 he then moved (u1/5l) left ….and so on until he had completed the square shown below:

75	70	41	12	52	59	30	7	23	369
10	51	62	28	6	26	73	69	44	369
31	9	20	76	72	38	13	54	56	369
77	66	43	14	48	61	32	3	25	369
17	46	60	35	1	24	80	64	42	369
29	4	27	74	67	45	11	49	63	369
79	68	39	16	50	57	34	5	21	369
15	53	55	33	8	19	78	71	37	369
36	2	22	81	65	40	18	47	58	369
369	369	369	369	369	369	369	369	369	

Fig 20

His treatment of triplets intrigued me, so I listed them:

1	2	159	168		
2	3	249	258	267	
3	2	348	357		
4	3	429	438	456	
5	4	519	528	537	546
6	3	618	627	654	
7	2	726	753		
8	3	816	843	852	
9	2	924	951		
	24				

Fig 21

I could now see how he derived his figure of 24^2 and it became clear to me that any pandiagonal Magic Squares derived from this method would have to contain three of the above triplets in every row and column.

With all the data he had provided I set about compiling the 576 pandiagonal Magic Squares of the 9th order he thought possible. I succeeded in compiling only 256 from the following list of 16 interchangeable sequence changers.

Kelsey	123 564 978	123 845 697	
	123 568 974	123 847 695	
	123 574 968	123 895 647	Leeflang
	123 578 964	123 897 645	
	123 645 897	123 964 578	
	123 647 895	123 968 574	Kelsey
	123 695 847	123 974 568	
Leeflang	123 697 845	123 978 564	

(The groups highlighted in yellow are the ones chosen by Leeflang, the ones in blue are those chosen by me and explained later.)

Intrigued to discover that I had obviously missed eight groups, I did a number frequency count which disclosed the missing groups:

123 456 789
123 459 786
123 486 759
123 489 756
123 756 489
123 759 486
123 786 459
123 789 456

It immediately became clear why these groups could not produce the required squares. They were not compatible with triplets. The figures in the 4th column are either 4 or 7. A triplet containing either of them and the figure 1 would have to be 1,4,10 or 1,7,7, obvious impossibilities.

Let's look at the Lo Shu square again and consider the eight triplets adding up to 15, i.e. the 3 rows and columns and the 2 diagonals.

2	9	4
7	5	3
6	1	8

It is significant that the 16 sets of viable triplets listed previously are all in linear relationship, while the 8 failed sets are not. **Fig 19** shows how these sets of triplets are used in pairs to create pandiagonal Magic Squares, but since the two diagonals of the Lo Shu square contain the common number 5, only six of the eight pairings are possible.

For his example Leeflang chose the numbers in the *rows* of the Lo Shu square to compile his pairings. I wondered if choosing the numbers in the *columns* would produce a different outcome or merely be a re-arrangement of the same Magic Square, so I set out to find the answer.
I chose the groups highlighted blue above as they contained the numbers in the Lo Shu *columns*.
I followed Leeflang's steps assiduously, including his route through the second square in order to make any comparison transparent, and produced the following:

1	2	3	4	5	6	7	8	9		1		1	2	3	9	6	8	5	7	4
10	11	12	13	14	15	16	17	18		2		10	11	12	18	15	17	14	16	13
19	20	21	22	23	24	25	26	27		3		19	20	21	27	24	26	23	25	22
28	29	30	31	32	33	34	35	36		5		37	38	39	45	42	44	41	43	40
37	38	39	40	41	42	43	44	45		6		46	47	48	54	51	53	5	52	49
46	47	48	49	50	51	52	53	54		4		28	29	30	36	33	35	32	34	31
55	56	57	58	59	60	61	62	63		9		73	74	74	81	78	80	77	79	76
64	65	66	67	68	69	70	71	72		7		55	56	57	63	60	62	59	61	58
73	74	74	76	77	78	79	80	81		8		64	65	66	72	69	71	68	70	67

Fig 22

This led to the creation of the following pandiagonal 9x9 Magic Square.

48	76	71	12	40	35	57	4	26	369
10	41	36	55	5	27	46	77	72	369
61	6	20	52	78	65	16	42	29	369
53	75	67	17	39	31	62	3	22	369
18	37	32	63	1	23	54	73	68	369
56	7	24	47	79	69	11	43	33	369
49	80	66	13	44	30	58	8	21	369
14	45	28	59	9	19	50	81	64	369
60	2	25	51	74	70	15	38	34	369
369	369	369	369	369	369	369	369	369	

Fig 23

This is a different Magic Square from Leeflang's, but upon close examination I found that 9 numbers occupied the same position in both his and my squares.

75	70	41	12	52	59	30	7	23
10	51	62	28	6	26	73	69	44
31	9	20	76	72	38	13	54	56
77	66	43	14	48	61	32	3	25
17	46	60	35	1	24	80	64	42
29	4	27	74	67	45	11	49	63
79	68	39	16	50	57	34	5	21
15	53	55	33	8	19	78	71	37
36	2	22	81	65	40	18	47	58

48	76	71	12	40	35	57	4	26
10	41	36	55	5	27	46	77	72
61	6	20	52	78	65	16	42	29
53	75	67	17	39	31	62	3	22
18	37	32	63	1	23	54	73	68
56	7	24	47	79	69	11	43	33
49	80	66	13	44	30	58	8	21
14	45	28	59	9	19	50	81	64
60	2	25	51	74	70	15	38	34

Fig 24

The duplicated numbers are :

1	2	3
10	11	12
19	20	21

which re-arranged in Lo Shu order becomes

2	21	10	33
19	11	3	33
12	1	20	33
33	33	33	

Whether this is a quirky coincidence or mathematically obvious I leave you to judge.
An examination of both 9x9 squares reveals that they are pandiagonal, the rows, columns, diagonals and broken diagonals all total 369, and both squares contain 81 subsets of 3x3 squares, giving a total of 117 constants for each square.

Constant Count

Rows	9
Columns	9
Fwd Diagonals	9
Bwd Diagonals	9
Subsets of 3x3 cells	81
	_____
Total	117
	=====

There are no more subsets available so both squares are perfect.

* * *

Puzzle Corner

This square contains the numbers 1 to 81 in random order with some numbers not shown. When the missing numbers and their positions are identified the total of all rows, columns and diagonals and broken diagonals will add up to 369 as will the contents of the nine smaller squares within its borders as well as any subset of nine numbers which together form a square.

37	53		19		78	55	71	15
	9	74	58	72	11	40	54	
61		14			32	25	3	77
42	46	35	24	1			64	17
20	4		56	67	18	38		36
	70	12	41	52		23	7	75
44	51			6	73		69	
27	2	76	63		13	45	47	31
57		16	39	50	34	21		79

Missing numbers			
5	8	10	22
26	28	29	30
33	43	48	49
59	60	62	65
66	68	80	81

Puzzle No. 6

BOOK TWO

MAGIC CUBES

Preamble. I earlier said that much has been written about Magic Squares and their properties, but even more has been written about Magic Cubes. Once again my intention is to simplify things for the beginner, so I will limit my criteria.

The constant in Magic Cubes can be produced (a) by the planar squares' rows, columns, forwards facing and backwards facing diagonals and broken diagonals; (b) by the cubes' pillars and cubic diagonals, i.e. those passing through the cube; and (c) by subsets, non-linear groupings of cells which produce the constant as a direct result of the construction of the Magic Cube, not fortuitously.

Further, a cube can be viewed and assessed through three dimensions – from the front, from the side and from the top, which I will refer to as the X, Y and Z aspects..

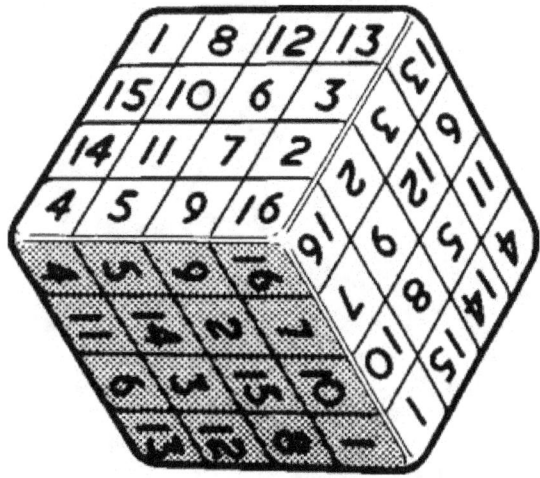

For the sake of simplicity I will confine the number of categories for Magic Cubes to three.

My three categories are:

Basic: where from one aspect all planar squares are pandiagonal but neither the pillars nor the cubic diagonals produce the constant.

Near perfect: where from one aspect all planar squares are pandiagonal and either all pillars or all cubic diagonals produce the constant, and where from the other two aspects no more than one grouping fails, be it row, column, pillar or cubic diagonal.

Perfect: where from one aspect all planar squares are pandiagonal and all pillars and all cubic diagonals produce the constant, and where from the other two aspects no more than one grouping fails, be it row, column, pillar or cubic diagonal.

As the constant is the common feature of all Magic Cubes I will use the number of ways a cube produces the constant as the means of comparison between cubes, and that detail will be shown in 'Constant Count' boxes after each cube as appropriate.

Planar constants are readily recognisable but cubic diagonals are not. Those wishing to follow the cubic pathways of subsequent Magic Cubes can do so via the Appendices at the end of this book. As it is impossible for the cubic diagonals of all the cells in planar square 1 to proceed unbroken to their destination in planar square ŋ, the pathways detailed contain broken diagonals. The unbroken diagonals are shown separately.

* * *

3x3 Magic Cubes

Any Magic Square can be converted to a Magic Cube by a method which, in theory, is very simple. Let us start with a 3x3 Magic Cube. First we set out the basic square and create two other squares by adding 9 and 18 consecutively to these further squares:

2	9	4
7	5	3
6	1	8

11	18	13
16	14	12
15	10	17

20	27	22
25	23	21
24	19	26

The numbers 1 to 27 must now be spread equally among the three squares so that each square contains three numbers from its two companions. We start by repositioning the column numbers in the first square in sequence to form the first row of all three squares:

2	7	6

9	5	1

4	3	8

The blank cells are then filled by inserting numbers from the sister squares in sequence:

2	7	6	
16	15	11	42
24	20	25	
42	42	42	

9	5	1	
14	10	18	42
19	27	23	
42	42	42	

4	3	8	
12	17	13	42
26	22	21	
42	42	42	

Fig 25

You can see that the columns all give the constant of 42 as do the centre rows of the three squares and their backwards facing diagonals . What is not so obvious is the fact that two cubic diagonals also produce the constant: 24, 10 and 8; plus 6, 10 and 26. Further, the numbers in the middle row of each square, viewed as pillars of the cube, 16, 14 and 12; 15, 10 and 17; plus 11, 18 and 13 also sum up to the constant.

Versatile though this cube is, it is far from perfect. Not one of the three squares is a Magic Square as two of its rows fail to produce the constant; none of the forward facing diagonals work; and only three of the nine pillars succeed.

So how do we rate this cube?

Constant count

Columns	9
Rows	3
Planar diagonals	3
Cubic diagonals	2
Pillars	3
Total	20

The above method of constructing a Magic Cube will be explained more fully in the Chapter on 5x5 cubes, and there are other methods which we will explore as we progress.

* * *

4x4 Magic Cubes

It has been proved that a perfect 4x4 Magic Cube is mathematically impossible, so in this Chapter we will strive to construct a cube as near perfect as possible. For the constant of any Magic Cube to be produced by cubic diagonals, the original Magic Square must be pandiagonal.

Using the method I outlined in the previous chapter, I constructed the following cube by creating four pandiagonal Magic Squares, spreading the numbers 1 to 64 throughout, to produce a constant of 130 in each of the cube's three aspects, that is, when the cube is viewed from the top, from the front and from the side, which I call the X, Y and Z aspects.

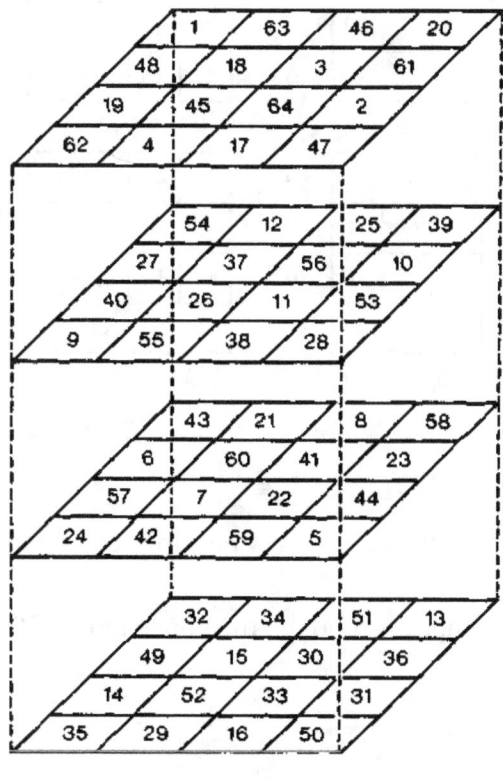

Fig 26

I have depicted the cube in the above 3D form to illustrate its transition from square to cube, but it is deficient in displaying the whole picture. It shows only the view from the top of the cube. The cube can be more clearly examined by showing all squares in all aspects as follows:

X aspect

1	63	46	20	130
48	18	3	61	130
19	45	64	2	130
62	4	17	47	130
130	130	130	130	

54	12	25	39	130
27	37	56	10	130
40	26	11	53	130
9	55	38	28	130
130	130	130	130	

43	21	8	58	130
6	60	41	23	130
57	7	22	44	130
24	42	59	5	130
130	130	130	130	

32	34	51	13	130
49	15	30	36	130
14	52	33	31	130
35	29	16	50	130
130	130	130	130	

Y aspect

1	48	19	62	130
54	27	40	9	130
43	6	57	24	130
32	49	14	35	130
130	130	130	130	

63	18	45	4	130
12	37	26	55	130
21	60	7	42	130
34	15	52	29	130
130	130	130	130	

46	3	64	17	130
25	56	11	38	130
8	41	22	59	130
51	30	33	16	130
130	130	130	130	

20	61	2	47	130
39	10	53	28	130
58	23	44	5	130
13	36	31	50	130
130	130	130	130	

Z aspect

1	54	43	32	130
63	12	21	34	130
46	25	8	51	130
20	39	58	13	130
130	130	130	130	

48	27	6	49	130
18	37	60	15	130
3	56	41	30	130
61	10	23	36	130
130	130	130	130	

19	40	57	14	130
45	26	7	52	130
64	11	22	33	130
2	53	44	31	130
130	130	130	130	

62	9	24	35	130
4	55	42	29	130
17	38	59	16	130
47	28	5	50	130
130	130	130	130	

Fig 27

An interesting grouping is that of subsets, namely any four cells which together form a square.

In this cube the following groupings all produce the constant of 130.

Constant Count

	X aspect	Y aspect	Z aspect	Total
Rows	16	16	16	48
Columns	16	16	16	48
Fwd Diagonals	16			16
Bwd Diagonals	16			16
Subsets	64	20	24	108
Pillars	16	16	16	48
Cubic Diagonals				
Total	144	68	72	284

I found it disappointing that this cube produced no cubic diagonal constants, so I tried again with a different square:

r

X aspect

1	57	56	16	130
48	24	25	33	130
28	36	45	21	130
53	13	4	60	130
130	130	130	130	

62	6	11	51	130
19	43	38	30	130
34	26	23	47	130
15	55	58	2	130
130	130	130	130	

63	7	10	50	130
18	42	39	31	130
35	27	22	46	130
14	54	59	3	130
130	130	130	130	

5	61	52	12	130
44	20	29	37	130
32	40	41	17	130
49	9	8	64	130
130	130	130	130	

Y aspect

1	48	28	53	130
62	19	34	15	130
63	18	35	14	130
5	44	32	49	130
131	129	129	131	

57	24	36	13	130
6	43	26	55	130
7	42	27	54	130
61	20	40	9	130
131	129	129	131	

56	25	45	4	130
11	38	23	58	130
10	39	22	59	130
52	29	41	8	130
129	131	131	129	

16	33	21	60	130
51	30	47	2	130
50	31	46	3	130
12	37	17	64	130
129	131	131	129	

Z aspect

1	62	63	5	131
57	6	7	61	131
56	11	10	52	129
16	51	50	12	129
130	130	130	130	

48	19	18	44	129
24	43	42	20	129
25	38	39	29	131
33	30	31	37	131
130	130	130	130	

28	34	35	32	129
36	26	27	40	129
45	23	22	41	131
21	47	46	17	131
130	130	130	130	

53	15	14	49	131
13	55	54	9	131
4	58	59	8	129
60	2	3	64	129
130	130	130	130	

Fig 28

Although this produced a cube with successful cubic diagonals from all three aspects it is far from perfect as only one aspect has full pandiagonal characteristics. However, it does possess some remarkable characteristics. Squares 1 and 4 together and squares 2 and 3 together, in all three aspects, can be rotated through 90, 180 and 270 degrees without destroying the cubic diagonals. Does this enhance the cube's score, or am I merely pointing out that there can be fifteen very similar cubes?

Constant Count

	X aspect	Y aspect	Z aspect	Total
Rows	16	16		32
Columns	16		16	32
Fwd Diagonals	8			8
Bwd Diagonals	8			8
Subsets	32	16	8	56
Pillars		16	16	32
Cubic Diagonals	4	4	4	12
Total	84	52	44	180

How do we measure the relative perfection of the two cubes above? If the number of constants was the measuring stick then the second cube is less perfect than the first, even though it achieved all four cubic diagonals in all three aspects compared with the other cube's nil, and was capable of being morphed into fifteen similar cubes.

While I was drafting this I wondered - if the opposing squares were *counter* - rotated would the diagonal constants survive? I rotated squares 1 and 4 clockwise 90 degrees and squares 2 and 3 anticlockwise 90 degrees, (in other words performing one of the fifteen morphings available) and discovered that the cubic diagonals remained unaffected through all three aspects. I found that most amazing. Had I discovered a new feature of Magic Cubes? The morphed cube is shown below. APPENDIX VII sets out all sixteen morphings.

50

53	28	48	1	130		51	30	47	2	130		50	31	46	3	130		49	32	44	5	130	
13	36	24	57	130		11	38	23	58	130		10	39	22	59	130		9	40	20	61	130	X aspect
4	45	25	56	130		6	43	26	55	130		7	42	27	54	130		8	41	29	52	130	
60	21	33	16	130		62	19	34	15	130		63	18	35	14	130		64	17	37	12	130	
130	130	130	130			130	130	130	130			130	130	130	130			130	130	130	130		

5	63	62	1	131		13	55	54	9	131		4	58	59	8	129		12	50	51	16	129	
44	18	19	48	129		36	26	27	40	129		45	23	22	41	131		37	31	30	33	131	Y aspect
32	35	34	28	129		24	43	42	20	129		25	38	39	29	131		17	46	47	21	131	
49	14	15	53	131		57	6	7	61	131		56	11	10	52	129		64	3	2	60	129	
130	130	130	130			130	130	130	130			130	130	130	130			130	130	130	130		

16	56	57	1	130		44	20	29	37	130		32	40	41	17	130		60	4	13	53	130	
51	11	6	62	130		18	42	39	31	130		35	27	22	46	130		2	58	55	15	130	Z aspect
50	10	7	63	130		19	43	38	30	130		34	26	23	47	130		3	59	54	14	130	
12	52	61	5	130		48	24	25	33	130		28	36	45	21	130		64	8	9	49	130	
129	129	131	131			129	129	131	131			129	129	131	131			129	129	131	131		

Fig 29

This Magic Cube's Constant Count is less than the un-morphed version's because of the loss of all pillar constants.

Constant Count

	X aspect	Y aspect	Z aspect	Total
Rows	16		16	32
Columns	16	16		32
Fwd Diagonals	16			16
Bwd Diagonals	16			16
Subsets	20	16	8	44
Pillars	0	0	0	0
Cubic Diagonals	16	16	16	48
	___	___	___	___
Total	100	48	40	188
	=====	=====	=====	=====

Even so, it was a pleasant surprise to find counter-rotating possibilities with Magic Squares and Cubes. How wonderful it would be to create a counter–rotating puzzle.

Puzzle Corner

I'd like to tell you about an eccentric inventor who bequeathed to his son a patent specification relating to a combination mechanism for safes. It comprised 32 tumblers numbered from 1 to 32. When the combination was first set these tumblers had to be inserted into 4 cogwheels in such a way that the numbers in each wheel totalled 132. In addition every horizontal row of eight tumblers had also to total 132 and continue to do so even when the cogwheels were turned.

When the inventor originally filed his patent application he deliberately omitted several of the numbers from the specification drawings in order to keep the relative position of the tumblers a trade secret. He kept the completed drawing securely locked away in his private safe together with all his securities. When he died his son had a problem. He could not open the safe without the completed drawing and had only the patent application drawing to work on.

From the incomplete specification drawing shown below could you open the safe? In solving this puzzle bear in mind that there are 16 positions for each cogwheel, not 8 as one might initially believe. The constant of 132 is obtained from the horizontal lines of 8 tumblers in all 16 positions as well as the 8 tumblers in each cogwheel.

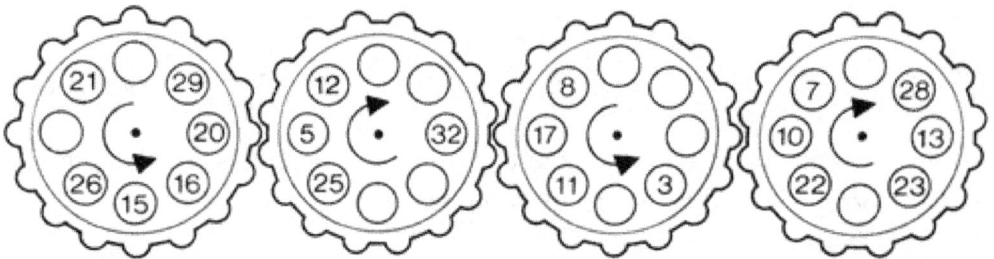

Puzzle No. 7

Well, I did tell you Magic Squares need not be square.

5x5 Magic Cubes

In the first square below I detail a pandiagonal Magic Square and have added 25 in succession to form four other squares.

24	12	5	18	6
3	16	9	22	15
7	25	13	1	19
11	4	17	10	23
20	8	21	14	2

65 65 65 65 65

49	37	30	43	31
28	41	34	47	40
32	50	38	26	44
36	29	42	35	48
45	33	46	39	27

190 190 190 190 190

74	62	55	68	56
53	66	59	72	65
57	75	63	51	69
61	54	67	60	73
70	58	71	64	52

315 315 315 315 315

99	87	80	93	81
78	91	84	97	90
82	100	88	76	94
86	79	92	85	98
95	83	96	89	77

440 440 440 440 440

124	112	105	118	106
103	116	109	122	115
107	125	113	101	119
111	104	117	110	123
120	108	121	114	102

565 565 565 565 565

Fig 30

I re-distributed the numbers of these 'old' squares to the new squares in the following sequence:
The rows of new **Square 1** were taken from the corresponding rows of the old squares in the sequence row by row of 1,2,3,4,5; 2,3,4,5,1; 3,4,5,1,2; 4,5,1,2,3 and 5,1,2,3,4.
The rows of new **Square 2** were taken from the corresponding rows of the old squares in the same sequence as above but with the count beginning with old square 2.
The rows of new **Square 3** were taken from the corresponding rows of the old squares in the same sequence as above but with the count beginning with old square 3.
and so on to produce the following 5x5 Magic Cube.

24	37	55	93	106	315
28	66	84	122	15	315
57	100	113	1	44	315
86	104	17	35	73	315
120	8	46	64	77	315
315	315	315	315	315	

49	62	80	118	6	315
53	91	109	22	40	315
82	125	13	26	69	315
111	4	42	60	98	315
20	33	71	89	102	315
315	315	315	315	315	

74	87	105	18	31	315
78	116	9	47	65	315
107	25	38	51	94	315
11	29	67	85	123	315
45	58	96	114	2	315
315	315	315	315	315	

99	112	5	43	56	315
103	16	34	72	90	315
7	50	63	76	119	315
36	54	92	110	23	315
70	83	121	14	27	315
315	315	315	315	315	

124	12	30	68	81	315
3	41	59	97	115	315
32	75	88	101	19	315
61	79	117	10	48	315
95	108	21	39	52	315
315	315	315	315	315	

X aspect

Fig 31 X aspect

24	28	57	86	120	315
49	53	82	111	20	315
74	78	107	11	45	315
99	103	7	36	70	315
124	3	32	61	95	315
370	265	285	305	350	

37	66	100	104	8	315
62	91	125	4	33	315
87	116	25	29	58	315
112	16	50	54	83	315
12	41	75	79	108	315
310	330	375	270	290	

55	84	113	17	46	315
80	109	13	42	71	315
105	9	38	67	96	315
5	34	63	92	121	315
30	59	88	117	21	315
275	295	315	335	355	

93	122	1	35	64	315
118	22	26	60	89	315
18	47	51	85	114	315
43	72	76	110	14	315
68	97	101	10	39	315
340	360	255	300	320	

106	15	44	73	77	315
6	40	69	98	102	315
31	65	94	123	2	315
56	90	119	23	27	315
81	115	19	48	52	315
280	325	345	365	260	

Y aspect

Fig 31 Y aspect

56

24	49	74	99	124	370
37	62	87	112	12	310
55	80	105	5	30	275
93	118	18	43	68	340
106	6	31	56	81	280
315	315	315	315	315	

28	53	78	103	3	265
66	91	116	16	41	330
84	109	9	34	59	295
122	22	47	72	97	360
15	40	65	90	115	325
315	315	315	315	315	

57	82	107	7	32	285
100	125	25	50	75	375
113	13	38	63	88	315
1	26	51	76	101	255
44	69	94	119	19	345
315	315	315	315	315	

86	111	11	36	61	305
104	4	29	54	79	270
17	42	67	92	117	335
35	60	85	110	10	300
73	98	123	23	48	365
315	315	315	315	315	

120	20	45	70	95	350
8	33	58	83	108	290
46	71	96	121	21	355
64	89	114	14	39	320
77	102	2	27	52	260
315	315	315	315	315	

Z aspect

Fig 31 Z aspect

The subsets in this cube are any five numbers in cross formation, either upright or slanting, which produce the constant.

Constant Count

	X aspect	Y aspect	Z aspect	Total
Rows	25	25		50
Columns	25		25	50
Fwd Diagonals				
Bwd Diagonals	25	25	25	75
Subsets	150	150	150	450
Pillars	25	25		50
Cubic Diagonals	25	25	25	75
Total	275	250	225	750

But the cube is far from perfect. None of the squares is pandiagonal and all the forward facing diagonals fail.

Let us now consider a different method of creating a Magic Cube, a method I call **the successive additive method**. From a different pandiagonal Magic Square let us construct the cube applying simple logic and thereby controlling the outcome. We saw that the original square had a constant of 65 which was augmented to 315 in the cube by successive additions of 25, 50, 75 and 100.

We begin with three 5x5 grids: the first being the pandiagonal Magic Square; the second being a Latin square with the letters a to e; and the third with the letters transformed to the additives 0, 25, 50, 75 and 100. This third square I call the **additives table**

24	12	5	18	6	65
3	16	9	22	15	65
7	25	13	1	19	65
11	4	17	10	23	65
20	8	21	14	2	65
65	65	65	65	65	

a	c	e	b	d
b	d	a	c	e
c	e	b	d	a
d	a	c	e	b
e	b	d	a	c

a	0
b	25
c	50
d	75
e	100

0	50	100	25	75	250
25	75	0	50	100	250
50	100	25	75	0	250
75	0	50	100	25	250
100	25	75	0	50	250
250	250	250	250	250	

Fig 32

Applying the additives table in sequence to the Magic Square produces a constant of 315 through the five layers of the cube.

I constructed the first planar square of the cube by adding the first row of the additives table to the first row of the square; the second additives row to the second planar row, and so on:

24	62	105	43	81
28	91	9	72	115
57	125	38	76	19
86	4	67	110	48
120	33	96	14	52

Fig 33

The other squares were constructed by moving the additives table one row down and one column to the right and applying it successively to the pandiagonal Magic Square, in effect applying five different additive tables in sequence to the pandiagonal square.

0	50	100	25	75
25	75	0	50	100
50	100	25	75	0
75	0	50	100	25
100	25	75	0	50

75	0	50	100	25
100	25	75	0	50
0	50	100	25	75
25	75	0	50	100
50	100	25	75	0

25	75	0	50	100
50	100	25	75	0
75	0	50	100	25
100	25	75	0	50
0	50	100	25	75

100	25	75	0	50
0	50	100	25	75
25	75	0	50	100
50	100	25	75	0
75	0	50	100	25

50	100	25	75	0
75	0	50	100	25
100	25	75	0	50
0	50	100	25	75
25	75	0	50	100

This produced the X aspect of the 5x5 Magic Cube,

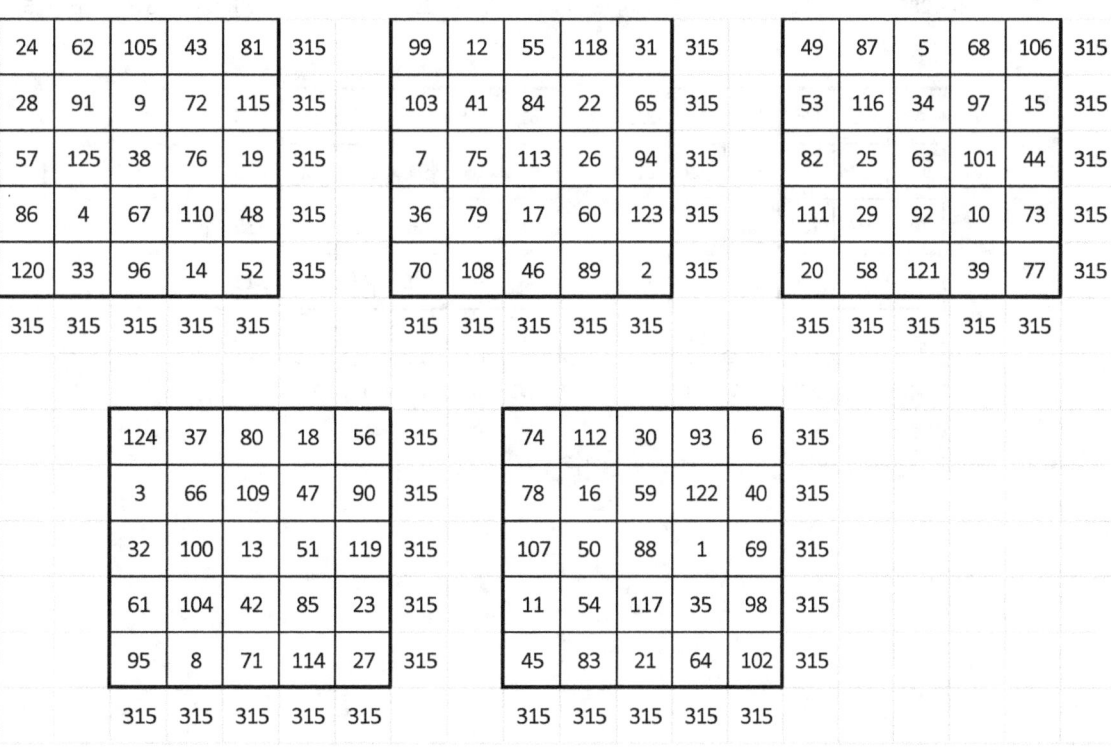

Fig 34 X aspect

from which I was able to compile the other two aspects.

24	28	57	86	120	315
99	103	7	36	70	315
49	53	82	111	20	315
124	3	32	61	95	315
74	78	107	11	45	315
370	265	285	305	350	

62	91	125	4	33	315
12	41	75	79	108	315
87	116	25	29	58	315
37	66	100	104	8	315
112	16	50	54	83	315
310	330	375	270	290	

105	9	38	67	96	315
55	84	113	17	46	315
5	34	63	92	121	315
80	109	13	42	71	315
30	59	88	117	21	315
275	295	315	335	355	

43	72	76	110	14	315
118	22	26	60	89	315
68	97	101	10	39	315
18	47	51	85	114	315
93	122	1	35	64	315
340	360	255	300	320	

81	115	19	48	52	315
31	65	94	123	2	315
106	15	44	73	77	315
56	90	119	23	27	315
6	40	69	98	102	315
280	325	345	365	260	

Fig 34 Y aspect

24	99	49	124	74	370
62	12	87	37	112	310
105	55	5	80	30	275
43	118	68	18	93	340
81	31	106	56	6	280
315	315	315	315	315	

28	103	53	3	78	265
91	41	116	66	16	330
9	84	34	109	59	295
72	22	97	47	122	360
115	65	15	90	40	325
315	315	315	315	315	

57	7	82	32	107	285
125	75	25	100	50	375
38	113	63	13	88	315
76	26	101	51	1	255
19	94	44	119	69	345
315	315	315	315	315	

86	36	111	61	11	305
4	79	29	104	54	270
67	17	92	42	117	335
110	60	10	85	35	300
48	123	73	23	98	365
315	315	315	315	315	

120	70	20	95	45	350
33	108	58	8	83	290
96	46	121	71	21	355
14	89	39	114	64	320
52	2	77	27	102	260
315	315	315	315	315	

Fig 34 Z aspect

The subsets in this cube are any five numbers in cross formation, either upright or slanting, which produce the constant.

Constant Count

	X aspect	Y aspect	Z aspect	Total
Rows	25	25		50
Columns	25		25	50
Fwd Diagonals	25	25	25	75
Bwd Diagonals	25	25	25	75
Upright Subsets	125	125	125	375
Slanting Subsets	125	125	125	375
Pillars		25	25	50
Cubic Diagonals	25	25	25	75
Total	375	375	375	1,125

Although each aspect suffers one failure of either pillars, rows or columns, this cube merits rating as perfect.

To the best of my knowledge and belief this is the first time a 5x5 Magic cube of this complexity has been published.

* * *

Puzzle Corner

When the ladies of the Oxbridge Entomological Society returned recently from their expedition to Central America they brought back a specimen of the hives of the Incredible Adding Bee (gen. apis arithmeticus) so named because of the precise numerical accuracy with which the Queen Bee deposits her eggs. She selects one empty hexagonal cell and in each of the six cells which surround it she deposits a different number of eggs. She continues this fashion until she has filled 36 cells all containing a different number of eggs, and then starts the process over again. The incredible aspect of this ritual, the purpose of which is not yet understood, is that the eggs in the cells surrounding each empty cell always total exactly 111, despite the fact that each occupied cell can form part of three such groupings.

When the specimen was unpacked it was discovered that the contents of several cells had perished in transit, but fortunately a sufficient number of cells survived intact to enable the ladies to ascertain the total number of eggs which were originally deposited in each cell.

From the following diagram of the specimen as it was unpacked, can you supply the missing data?

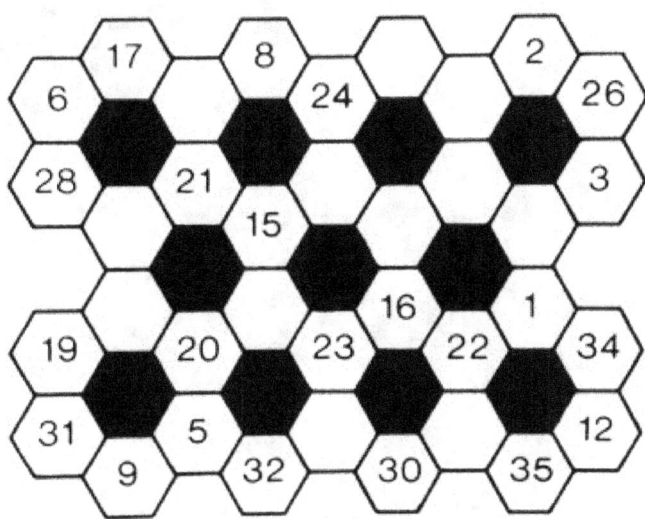

Puzzle No. 8

6x6 Magic Cubes

For a magic Cube to be successful it must originate from a pandiagonal Magic Square. Since no such thing exists in 6x6 form it might seem pointless to set out a 6x6 Magic Cube. Nonetheless, let's go through the procedure. From the square created in a previous chapter I have added 36 progressively to five other grids.

35	28	3	4	5	36
6	11	25	20	18	31
10	22	16	13	23	27
30	17	19	26	12	7
29	24	14	15	21	8
1	9	34	33	32	2

71	64	39	40	41	72
42	47	61	56	54	67
46	58	52	49	59	63
66	53	55	62	48	43
65	60	50	51	57	44
37	45	70	69	68	38

107	100	75	76	77	108
78	83	97	92	90	103
82	94	88	85	95	99
102	89	91	98	84	79
101	96	86	87	93	80
73	81	106	105	104	74

143	136	111	112	113	144
114	119	133	128	126	139
118	130	124	121	131	135
138	125	127	134	120	115
137	132	122	123	129	116
109	117	142	141	140	110

179	172	147	148	149	180
150	155	169	164	162	175
154	166	160	157	167	171
174	161	163	170	156	151
173	168	158	159	165	152
145	153	178	177	176	146

215	208	183	184	185	216
186	191	205	200	198	211
190	202	196	193	203	207
210	197	199	206	192	187
209	204	194	195	201	188
181	189	214	213	212	182

Fig 35

I next created the first column of the first square by a figure from the first column of the other squares in turn. The second column of the first square was formed by the figures from the second column of the other squares, and so on until the square was completed. The second square was created in the same way, beginning with the figure heading the second column of the first square. Similarly the third square was started with the figure heading the third column of the first square, and so on. The ensuing Magic Cube is as follows:

35	64	75	112	149	216	651
42	83	133	164	198	31	651
82	130	160	193	23	63	651
138	161	199	26	48	79	651
173	204	14	51	93	116	651
181	9	70	105	140	146	651
651	651	651	651	651	651	

28	39	76	113	180	215	651
47	97	128	162	211	6	651
94	124	157	203	27	46	651
125	163	206	12	43	102	651
168	194	15	57	80	137	651
189	34	69	104	110	145	651
651	651	651	651	651	651	

3	40	77	144	179	208	651
61	92	126	175	186	11	651
88	121	167	207	10	58	651
127	170	192	7	66	89	651
158	195	21	44	101	132	651
214	33	68	74	109	153	651
651	651	651	651	651	651	

4	41	108	143	172	183	651
56	90	139	150	191	25	651
85	131	171	190	22	52	651
134	156	187	30	53	91	651
159	201	8	65	96	122	651
213	32	38	73	117	178	651
651	651	651	651	651	651	

5	72	107	136	147	184	651
54	103	114	155	205	20	651
95	135	154	202	16	49	651
120	151	210	17	55	98	651
165	188	29	60	86	123	651
212	2	37	81	142	177	651
651	651	651	651	651	651	

36	71	100	111	148	185	651
67	78	119	169	200	18	651
99	118	166	196	13	59	651
115	174	197	19	62	84	651
152	209	24	50	87	129	651
182	1	45	106	141	176	651
651	651	651	651	651	651	

Fig 36 X aspect

35	42	82	138	173	181	651
28	47	94	125	168	189	651
3	61	88	127	158	214	651
4	56	85	134	159	213	651
5	54	95	120	165	212	651
36	67	99	115	152	182	651
111	327	543	759	975	1191	

64	83	130	161	204	9	651
39	97	124	163	194	34	651
40	92	121	170	195	33	651
41	90	131	156	201	32	651
72	103	135	151	188	2	651
71	78	118	174	209	1	651
327	543	759	975	1191	111	

75	133	160	199	14	70	651
76	128	157	206	15	69	651
77	126	167	192	21	68	651
108	139	171	187	8	38	651
107	114	154	210	29	37	651
100	119	166	197	24	45	651
543	759	975	1191	111	327	

112	164	193	26	51	105	651
113	162	203	12	57	104	651
144	175	207	7	44	74	651
143	150	190	30	65	73	651
136	155	202	17	60	81	651
111	169	196	19	50	106	651
759	975	1191	111	327	543	

149	198	23	48	93	140	651
180	211	27	43	80	110	651
179	186	10	66	101	109	651
172	191	22	5	96	117	603
147	205	16	55	86	142	651
148	200	13	62	87	141	651
975	1191	111	279	543	759	

216	31	63	79	116	146	651
215	6	46	102	137	145	651
208	11	58	89	132	153	651
183	25	52	91	122	178	651
184	20	49	98	123	177	651
185	18	59	84	129	176	651
1191	111	327	543	759	975	

Fig 36 Y aspect

138	125	127	134	120	115	759
161	163	170	156	11	174	835
199	206	192	187	210	197	1191
26	12	7	30	17	19	111
48	43	66	5	55	62	279
79	102	89	91	98	84	543
651	651	651	603	511	651	

173	168	158	159	165	152	975
204	194	195	201	188	209	1191
14	15	21	8	29	24	111
51	57	44	65	60	50	327
93	80	101	96	86	87	543
116	137	132	122	23	129	659
651	651	651	651	551	651	

181	189	214	213	212	182	1191
9	34	33	32	2	1	111
70	69	68	38	37	45	327
105	104	74	73	81	106	543
140	110	109	117	142	141	759
146	145	153	178	177	176	975
651	651	651	651	651	651	

35	28	3	4	5	36	111
64	39	40	41	72	71	327
75	76	77	108	107	100	543
112	113	144	143	136	111	759
149	180	179	172	147	148	975
216	215	208	183	184	185	1191
651	651	651	651	651	651	

42	47	61	56	54	67	327
83	97	92	90	103	78	543
133	128	126	139	114	119	759
164	162	175	150	155	169	975
198	211	186	191	205	200	1191
31	6	11	25	20	18	111
651	651	651	651	651	651	

82	94	88	85	95	99	543
130	124	121	131	135	118	759
160	157	167	171	154	166	975
193	203	207	190	202	196	1191
28	27	10	22	16	13	116
63	46	58	52	49	59	327
656	651	651	651	651	651	

Fig 36 Z aspect

The X aspect produced 36 constants in each of the rows and columns.
The Y aspect produced 36 constants in each of the rows and pillars.
The Z aspect produced 36 constants in each of the columns and pillars.

Not a single diagonal, cubic or otherwise. We have merely created a series of planar squares which do not even rank as Magic Squares. In opening this Chapter I said that the effort might seem pointless, but we have at least confirmed that pointlessness! That's hardly worth writing home about so let's proceed to the Puzzle Corner.

* * *

Puzzle Corner

In the following diagram the circumferences equate to the rows of a magic square, the radii to the columns and the spirals to the diagonals. You are required to supply missing numbers so that the constant is seen as 150. The numbers are not in a continuous sequence, being 36 of the numbers 1 to 49.

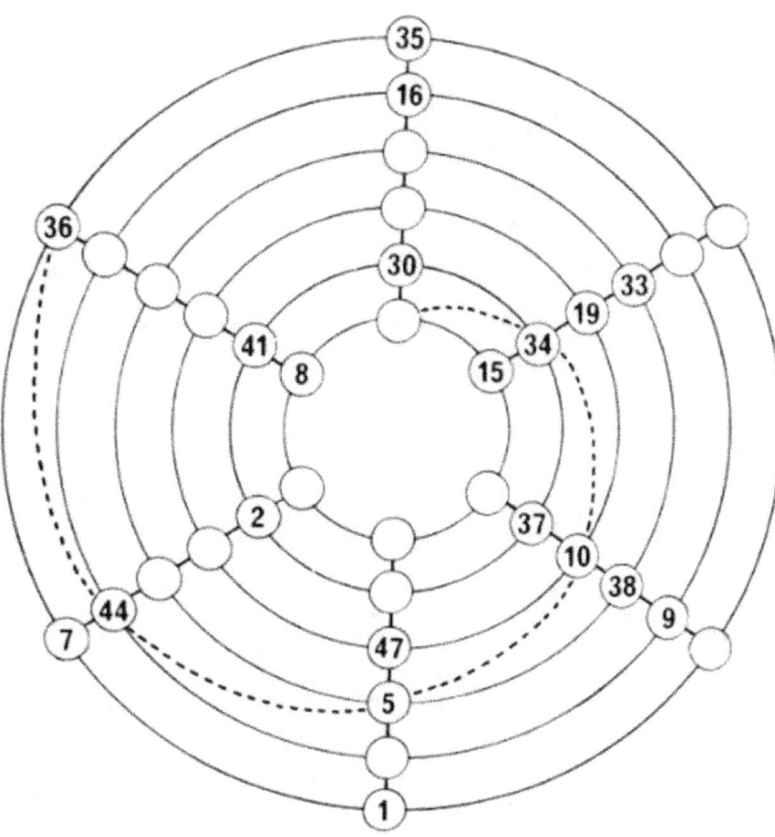

Puzzle No. 9

7x7 Magic Cubes

A few years ago a Magic Square complier uploaded a 7x7 Magic Cube which purported to possess the following properties: all squares were pandiagonal, and all rows, columns, pillars and cubic diagonals produced the constant of 1204. .

I was impressed by the skill of this unknown compiler and analysed the cube to ascertain the 7x7 pandiagonal square from which it was derived. I was in for a surprise as the square proved not to be pandiagonal, nor even a Magic Square.

45	24	3	31	10	38	17	168
35	14	42	21	49	28	7	196
18	46	25	4	32	11	39	175
1	29	8	36	15	43	22	154
40	19	47	26	5	33	12	182
23	2	30	9	37	16	44	161
13	41	20	48	27	6	34	189
175	175	175	175	175	175	175	

Fig 37

Knowing that all Magic Cubes must originate from a pandiagonal Magic Square I set out to find the fallacies contained in the cube. I soon found that the first planar square of the cube contained an error of 70 in one row and one column.

327	41	98	99	156	213	270	1204
52	109	166	223	280	330	44	1204
169	226	283	340	5	62	119	1204
293	301	8	297	122	17	236	1274
18	75	132	189	239	247	304	1204
135	192	200	25	314	260	78	1204
210	260	317	31	88	145	153	1204
1204	1204	1204	1204	1204	1274	1204	

Fig 38

However, the offending cell, the cell over-stated by 70, contained the number 17 making an over-statement impossible. The cube clearly contained other fallacies. I did a number count and found that four numbers, 28, 65, 179 and 257, totalling 529, were missing, It followed that four other numbers must have been duplicated and a search revealed them to be 29, 297, 17 and 260, totalling 599. The difference of 70 between the sum of the missing numbers and the sum of the duplicated numbers could not be hidden and was inevitably revealed in the first square. I detail below the cube I downloaded:

327	41	98	99	156	213	270
52	109	166	223	280	330	44
169	226	283	340	5	62	119
293	301	8	297	122	17	236
18	*75*	132	189	239	247	304
135	192	200	25	314	260	78
210	260	317	31	88	145	153

113	170	227	284	341	6	63
237	294	295	9	66	123	180
305	19	76	133	183	240	248
79	136	193	201	258	315	22
154	*204*	261	318	32	89	146
271	328	42	92	100	157	214
45	53	110	167	224	274	331

249	306	20	77	127	184	241
23	80	137	194	202	259	309
147	148	205	262	319	33	90
215	272	329	36	93	101	158
332	*46*	54	111	168	218	275
57	114	171	228	285	342	7
181	238	288	296	10	67	124

91	141	149	206	263	320	34
159	216	273	323	37	94	102
276	333	47	55	112	162	219
1	58	115	172	229	286	343
125	*182*	232	289	297	11	68
242	250	307	21	71	128	185
310	24	81	138	195	203	253

220	277	334	48	56	106	163
337	2	59	116	173	230	287
69	126	176	233	290	298	12
186	243	251	308	15	72	129
254	*311*	25	82	139	196	197
35	85	142	150	207	264	321
103	160	217	267	324	38	95

13	70	120	177	234	291	299
130	187	244	252	302	16	73
198	255	312	26	83	140	190
322	29	86	143	151	208	265
96	*104*	161	211	268	325	39
164	221	278	335	49	50	107
281	338	3	60	117	174	231

191	199	256	313	27	84	134
266	316	30	87	144	152	209
40	97	105	155	212	269	326
108	165	222	279	336	43	51
225	*282*	339	4	61	118	175
300	14	64	121	178	235	292
74	131	188	245	246	303	17

Fig 39

I have highlighted the offending numbers which could not have arisen unintentionally. I went back to the website to point out the fallacy of the cube but found that it had already been taken down. Some other enthusiast had had the same thoughts as me.

However, to return to the purposes of this book, I created the following Cube using the successive additives method from the pandiagonal Magic Square we produced earlier

1	18	35	45	13	23	40	175
24	41	2	19	29	46	14	175
47	8	25	42	3	20	30	175
21	31	48	9	26	36	4	175
37	5	15	32	49	10	27	175
11	28	38	6	16	33	43	175
34	44	12	22	39	7	17	175
175	175	175	175	175	175	175	

a	d	f	b	c	e	g	a	0
e	g	a	d	f	b	c	b	49
b	c	e	g	a	d	f	c	98
d	f	b	c	e	g	a	d	147
g	a	d	f	b	c	e	e	196
c	e	g	a	d	f	b	f	245
f	b	c	e	g	a	d	g	294

Fig 40

1	165	280	94	111	219	334	1204
220	335	2	166	274	95	112	1204
96	106	221	336	3	167	275	1204
168	276	97	107	222	330	4	1204
331	5	162	277	98	108	223	1204
109	224	332	6	163	278	92	1204
279	93	110	218	333	7	164	1204
1204	1204	1204	1204	1204	1204	1204	

84	143	209	317	40	148	263	1204
149	264	78	144	210	318	41	1204
319	42	150	265	79	145	204	1204
146	205	320	36	151	266	80	1204
260	81	147	206	321	37	152	1204
38	153	261	82	141	207	322	1204
208	316	39	154	262	83	142	1204
1204	1204	1204	1204	1204	1204	1204	

307	23	187	246	67	133	241	1204
127	242	308	24	188	247	68	1204
248	69	128	243	302	25	189	1204
26	183	249	70	129	244	303	1204
245	304	27	184	250	64	130	1204
65	131	239	305	28	185	251	1204
186	252	66	132	240	306	22	1204
1204	1204	1204	1204	1204	1204	1204	

285	50	116	231	339	13	170	1204
14	171	286	51	117	225	340	1204
226	341	8	172	287	52	118	1204
53	119	227	342	9	173	281	1204
174	282	54	113	228	343	10	1204
337	11	175	283	55	114	229	1204
115	230	338	12	169	284	56	1204
1204	1204	1204	1204	1204	1204	1204	

214	329	45	160	268	89	99	1204
90	100	215	323	46	161	269	1204
155	270	91	101	216	324	47	1204
325	48	156	271	85	102	217	1204
103	211	326	49	157	272	86	1204
273	87	104	212	327	43	158	1204
44	159	267	88	105	213	328	1204
1204	1204	1204	1204	1204	1204	1204	

192	258	72	138	197	312	35	1204
313	29	193	259	73	139	198	1204
140	199	314	30	194	253	74	1204
254	75	134	200	315	31	195	1204
32	196	255	76	135	201	309	1204
202	310	33	190	256	77	136	1204
71	137	203	311	34	191	257	1204
1204	1204	1204	1204	1204	1204	1204	

121	236	295	18	182	290	62	1204
291	63	122	237	296	19	176	1204
20	177	292	57	123	238	297	1204
232	298	21	178	293	58	124	1204
59	125	233	299	15	179	294	1204
180	288	60	126	234	300	16	1204
301	17	181	289	61	120	235	1204
1204	1204	1204	1204	1204	1204	1204	

Fig 41 X aspect

1	220	96	168	331	109	279	1204
84	149	319	146	260	38	208	1204
307	127	248	26	245	65	186	1204
285	14	226	53	174	337	115	1204
214	90	155	325	103	273	44	1204
192	313	140	254	32	202	71	1204
121	291	20	232	59	180	301	1204
1204	1204	1204	1204	1204	1204	1204	

165	335	106	276	5	224	93	1204
143	264	42	5	81	153	316	1004
23	242	69	183	304	131	252	1204
50	171	341	119	282	11	230	1204
329	100	270	48	211	87	159	1204
258	29	199	75	196	310	137	1204
236	63	177	298	125	288	17	1204
1204	1204	1204	1004	1204	1204	1204	

280	2	221	97	162	332	110	1204
209	78	150	320	147	261	39	1204
187	308	128	249	27	239	66	1204
116	286	8	227	54	175	338	1204
45	215	91	156	326	104	267	1204
72	193	314	134	255	33	203	1204
295	122	292	21	233	60	181	1204
1204	1204	1204	1204	1204	1204	1204	

94	166	336	107	277	6	218	1204
317	144	265	36	206	82	154	1204
246	24	243	70	184	305	132	1204
231	51	172	342	113	283	12	1204
160	323	101	271	49	212	88	1204
138	259	30	200	76	190	311	1204
18	237	57	178	299	126	289	1204
1204	1204	1204	1204	1204	1204	1204	

111	274	3	222	98	163	333	1204
40	210	79	151	321	141	262	1204
67	188	302	129	250	28	240	1204
339	117	287	9	228	55	169	1204
268	46	216	85	157	327	105	1204
197	73	194	315	135	256	34	1204
182	296	123	293	15	234	61	1204
1204	1204	1204	1204	1204	1204	1204	

219	95	167	330	108	278	7	1204
148	318	145	266	37	207	83	1204
133	247	25	244	64	185	306	1204
13	225	52	173	343	114	284	1204
89	161	324	102	272	43	213	1204
312	139	253	31	201	77	191	1204
290	19	238	58	179	300	120	1204
1204	1204	1204	1204	1204	1204	1204	

334	112	275	4	223	92	164	1204
263	41	204	80	152	322	142	1204
241	68	189	303	130	251	22	1204
170	340	118	281	10	229	56	1204
99	269	47	217	86	158	328	1204
35	198	74	195	309	136	257	1204
62	176	297	124	294	16	235	1204
1204	1204	1204	1204	1204	1204	1204	

Fig 41 Y aspect

76

1	84	307	285	214	192	121	1204
165	143	23	50	329	258	236	1204
280	209	187	116	45	72	295	1204
94	317	246	231	160	138	18	1204
111	40	67	339	268	197	182	1204
219	148	133	13	89	312	290	1204
334	263	241	170	99	35	62	1204
1204	1204	1204	1204	1204	1204	1204	.

220	149	127	14	90	313	291	1204
335	264	242	171	100	29	63	1204
2	78	308	286	215	193	122	1204
166	144	24	51	323	259	237	1204
274	210	188	117	46	73	296	1204
95	318	247	225	161	139	19	1204
112	41	68	340	269	198	176	1204
1204	1204	1204	1204	1204	1204	1204	.

96	319	248	226	155	140	20	1204
106	42	69	341	270	199	177	1204
221	150	128	8	91	314	292	1204
336	265	243	172	101	30	57	1204
3	79	302	287	216	194	123	1204
167	145	25	52	324	253	238	1204
275	204	189	118	47	74	297	1204
1204	1204	1204	1204	1204	1204	1204	

168	146	26	53	325	254	232	1204
276	205	183	119	48	75	298	1204
97	320	249	227	156	134	21	1204
107	36	70	342	271	200	178	1204
222	151	129	9	85	315	293	1204
330	266	244	173	102	31	58	1204
4	80	303	281	217	195	124	1204
1204	1204	1204	1204	1204	1204	1204	.

331	260	245	174	103	32	59	1204
5	81	304	282	211	196	125	1204
162	147	27	54	326	255	233	1204
277	206	184	113	49	76	299	1204
98	321	250	228	157	135	15	1204
108	37	64	343	272	201	179	1204
223	152	130	10	86	309	294	1204
1204	1204	1204	1204	1204	1204	1204	

109	38	65	337	273	202	180	1204
224	153	131	11	87	310	288	1204
332	261	239	175	104	33	60	1204
6	82	305	283	212	190	126	1204
163	141	28	55	327	256	234	1204
278	207	185	114	43	77	300	1204
92	322	251	229	158	136	16	1204
1204	1204	1204	1204	1204	1204	1204	

279	208	186	115	44	71	301	1204
93	316	252	230	159	137	17	1204
110	39	66	338	267	203	181	1204
218	154	132	12	88	311	289	1204
333	262	240	169	105	34	61	1204
7	83	306	284	213	191	120	1204
164	142	22	56	328	257	235	1204
1204	1204	1204	1204	1204	1204	1204	

Fig 41 Z aspect

The subsets in this cube are cells in 3-1-3 formation either horizontally or vertically.

This cube is perfect by any calculation.

Constant Count

	X aspect	Y aspect	Z aspect	Total
Rows	49	49	49	147
Columns	49	49	49	147
Fwd Diagonals	49	49	49	147
Bwd Diagonals	49		49	98
Horizontal Subsets	196	98	49	343
Vertical Subsets	98	49	49	196
Pillars	49	49	49	147
Cubic Diagonals	49	49	49	147
Total	588	392	392	1,372

To the best of my knowledge and belief this is the first time a 7x7 Magic cube of this complexity has been published.

* * *

Puzzle Corner

You are required to fill the blank cells with the missing numbers so as to complete a 7x7 pandiagonal square.

1	18	35	45	13	23	40	175
24	41	2		29	46	14	156
47	8				20	30	105
21						4	25
37	5				10	27	79
11	28	38		16	33	43	169
34	44	12	22	39	7	17	175
175	144	87	67	97	139	175	

Missing numbers

3	6	9
15	19	25
26	31	32
36	42	48
	49	

Puzzle No. 10

8x8 Magic Cubes

In this Chapter we will use one of the methods already outlined to create a pandiagonal Magic Square and then create two entirely different Magic Cubes from the same Magic Square.

We'll create the Magic Square by the Latin Square method using the two primary squares below.

A	F	G	D	C	H	E	B		a	b	g	h	e	f	c	d
G	D	A	F	E	B	C	H		b	a	h	g	f	e	d	c
B	E	H	C	D	G	F	A		h	g	b	a	d	c	f	e
H	C	B	E	F	A	D	G		g	h	a	b	c	d	e	f
B	E	H	C	D	G	F	A		c	d	e	f	g	h	a	b
H	C	B	E	F	A	D	G		d	c	f	e	h	g	b	a
A	F	G	D	C	H	D	B		f	e	d	c	b	a	h	g
G	D	A	F	E	B	C	H		e	f	c	d	a	b	g	h

Fig 42

If we apply the following values to the letters

A=1	a=0
B=2	b=8
C=3	c=16
D=4	d=24
E=5	e=32
F=6	f=40
G=7	g=48
H=8	h=56

we create an 8th order pandiagonal Magic Square with a constant of 260.

1	14	55	60	35	48	21	26	260
15	4	57	54	45	34	27	24	260
58	53	16	3	28	23	46	33	260
56	59	2	13	22	25	36	47	260
18	29	40	43	52	63	6	9	260
32	19	42	37	62	49	12	7	260
41	38	31	20	11	8	61	50	260
39	44	17	30	5	10	51	64	260
260	260	260	260	260	260	260	260	

Fig 43

We now need to apply a successive additives table to extrapolate the numbers 1 to 64 in the planar square into the numbers 1 to 512 in the 8x8 cube we are creating. We can use one of the primary squares in **Fig 36** with its values multiplied by 8.

a	b	g	h	e	f	c	d	a=0
b	a	h	g	f	e	d	c	b=64
h	g	b	a	d	c	f	e	c=128
g	h	a	b	c	d	e	f	d=192
c	d	e	f	g	h	a	b	e=256
d	c	f	e	h	g	b	a	f=320
f	e	d	c	b	a	h	g	g=384
e	f	c	d	a	b	g	h	h=448

0	64	384	448	256	320	128	192	1792
64	0	448	384	320	256	192	128	1792
448	384	64	0	192	128	320	256	1792
384	448	0	64	128	192	256	320	1792
128	192	256	320	384	448	0	64	1792
192	128	320	256	448	384	64	0	1792
320	256	192	128	64	0	448	384	1792
256	320	128	192	0	64	384	448	1792
1792	1792	1792	1792	1792	1792	1792	1792	

Fig 44

Applying the additives table gives us the following 8x8 Magic Cube:

1	78	439	508	291	368	149	218	2052
79	4	505	438	365	290	219	152	2052
506	437	80	3	220	151	366	289	2052
440	507	2	77	150	217	292	367	2052
146	221	296	363	436	511	6	73	2052
224	147	362	293	510	433	76	7	2052
361	294	223	148	75	8	509	434	2052
295	364	145	222	5	74	435	512	2052
2052	2052	2052	2052	2052	2052	2052	2052	

65	14	503	444	355	304	213	154	2052
463	388	121	54	237	162	347	280	2052
442	501	16	67	156	215	302	353	2052
184	251	258	333	406	473	36	111	2052
210	157	360	299	500	447	70	9	2052
352	275	234	165	126	49	460	391	2052
297	358	159	212	11	72	445	498	2052
39	108	401	478	261	330	179	256	2052
2052	2052	2052	2052	2052	2052	2052	2052	

449	398	119	60	227	176	341	282	2052
399	452	57	118	173	226	283	344	2052
186	245	272	323	412	471	46	97	2052
248	187	322	269	470	409	100	47	2052
338	285	232	171	116	63	454	393	2052
288	339	170	229	62	113	396	455	2052
41	102	415	468	267	328	189	242	2052
103	44	465	414	325	266	243	192	2052
2052	2052	2052	2052	2052	2052	2052	2052	

385	462	55	124	163	240	277	346	2052
143	196	313	374	429	482	27	88	2052
250	181	336	259	476	407	110	33	2052
376	315	194	141	86	25	484	431	2052
274	349	168	235	52	127	390	457	2052
32	83	426	485	318	369	140	199	2052
105	38	479	404	331	264	253	178	2052
487	428	81	30	197	138	371	320	2052
2052	2052	2052	2052	2052	2052	2052	2052	

129	206	311	380	419	496	21	90	2052
207	132	377	310	493	418	91	24	2052
378	309	208	131	92	23	494	417	2052
312	379	130	205	22	89	420	495	2052
18	93	424	491	308	383	134	201	2052
96	19	490	421	382	305	204	135	2052
489	422	95	20	203	136	381	306	2052
423	492	17	94	133	202	307	384	2052
2052	2052	2052	2052	2052	2052	2052	2052	

193	142	375	316	483	432	85	26	2052
335	260	249	182	109	34	475	408	2052
314	373	144	195	28	87	430	481	2052
56	123	386	461	278	345	164	239	2052
82	29	488	427	372	319	198	137	2052
480	403	106	37	254	177	332	263	2052
425	486	31	84	139	200	317	370	2052
167	236	273	350	389	458	51	128	2052
2052	2052	2052	2052	2052	2052	2052	2052	

321	270	247	188	99	48	469	410	2052
271	324	185	246	45	98	411	472	2052
58	117	400	451	284	343	174	225	2052
120	59	450	397	342	281	228	175	2052
466	413	104	43	244	191	326	265	2052
416	467	42	101	190	241	268	327	2052
169	230	287	340	395	456	61	114	2052
231	172	337	286	453	394	115	64	2052
2052	2052	2052	2052	2052	2052	2052	2052	

257	334	183	252	35	112	405	474	2052
15	68	441	502	301	354	155	216	2052
122	53	464	387	348	279	238	161	2052
504	443	66	13	214	153	356	303	2052
402	477	40	107	180	255	262	329	2052
160	211	298	357	446	497	12	71	2052
233	166	351	276	459	392	125	50	2052
359	300	209	158	69	10	499	448	2052
2052	2052	2052	2052	2052	2052	2052	2052	

Fig 45 X aspect

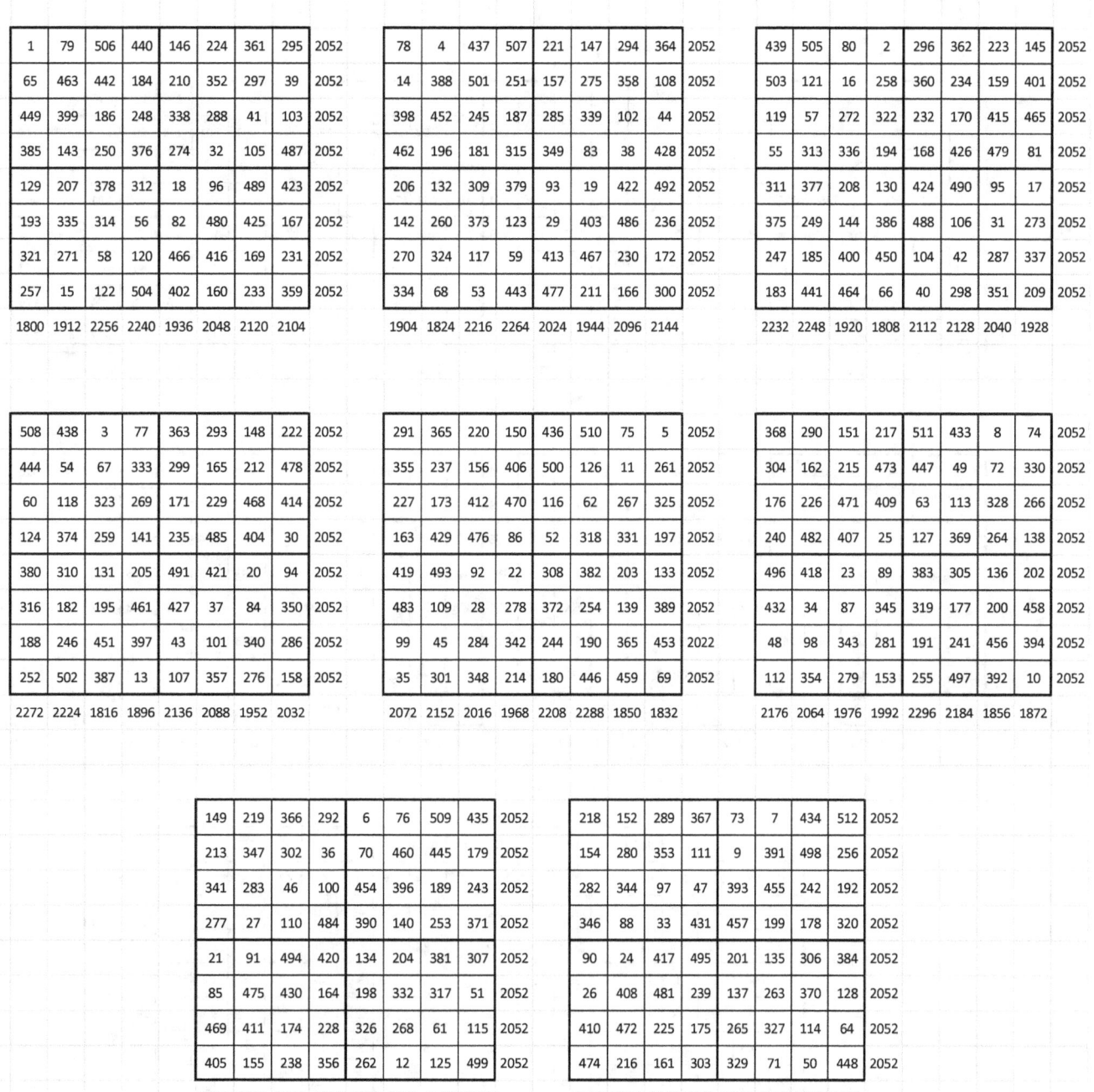

Fig 45 Y aspect

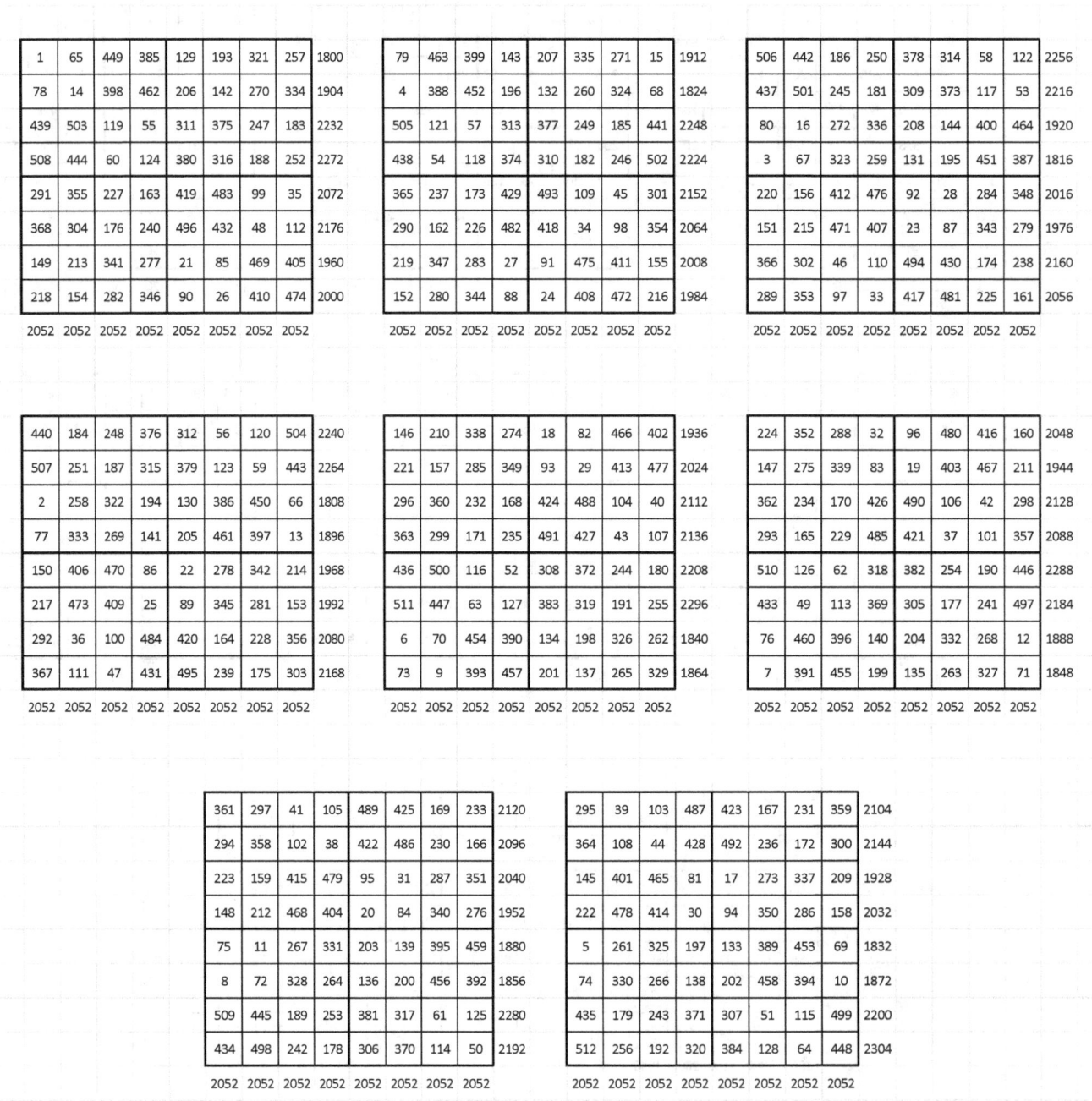

Table 1

1	65	449	385	129	193	321	257	1800
78	14	398	462	206	142	270	334	1904
439	503	119	55	311	375	247	183	2232
508	444	60	124	380	316	188	252	2272
291	355	227	163	419	483	99	35	2072
368	304	176	240	496	432	48	112	2176
149	213	341	277	21	85	469	405	1960
218	154	282	346	90	26	410	474	2000
2052	2052	2052	2052	2052	2052	2052	2052	

Table 2

79	463	399	143	207	335	271	15	1912
4	388	452	196	132	260	324	68	1824
505	121	57	313	377	249	185	441	2248
438	54	118	374	310	182	246	502	2224
365	237	173	429	493	109	45	301	2152
290	162	226	482	418	34	98	354	2064
219	347	283	27	91	475	411	155	2008
152	280	344	88	24	408	472	216	1984
2052	2052	2052	2052	2052	2052	2052	2052	

Table 3

506	442	186	250	378	314	58	122	2256
437	501	245	181	309	373	117	53	2216
80	16	272	336	208	144	400	464	1920
3	67	323	259	131	195	451	387	1816
220	156	412	476	92	28	284	348	2016
151	215	471	407	23	87	343	279	1976
366	302	46	110	494	430	174	238	2160
289	353	97	33	417	481	225	161	2056
2052	2052	2052	2052	2052	2052	2052	2052	

Table 4

440	184	248	376	312	56	120	504	2240
507	251	187	315	379	123	59	443	2264
2	258	322	194	130	386	450	66	1808
77	333	269	141	205	461	397	13	1896
150	406	470	86	22	278	342	214	1968
217	473	409	25	89	345	281	153	1992
292	36	100	484	420	164	228	356	2080
367	111	47	431	495	239	175	303	2168
2052	2052	2052	2052	2052	2052	2052	2052	

Table 5

146	210	338	274	18	82	466	402	1936
221	157	285	349	93	29	413	477	2024
296	360	232	168	424	488	104	40	2112
363	299	171	235	491	427	43	107	2136
436	500	116	52	308	372	244	180	2208
511	447	63	127	383	319	191	255	2296
6	70	454	390	134	198	326	262	1840
73	9	393	457	201	137	265	329	1864
2052	2052	2052	2052	2052	2052	2052	2052	

Table 6

224	352	288	32	96	480	416	160	2048
147	275	339	83	19	403	467	211	1944
362	234	170	426	490	106	42	298	2128
293	165	229	485	421	37	101	357	2088
510	126	62	318	382	254	190	446	2288
433	49	113	369	305	177	241	497	2184
76	460	396	140	204	332	268	12	1888
7	391	455	199	135	263	327	71	1848
2052	2052	2052	2052	2052	2052	2052	2052	

Table 7

361	297	41	105	489	425	169	233	2120
294	358	102	38	422	486	230	166	2096
223	159	415	479	95	31	287	351	2040
148	212	468	404	20	84	340	276	1952
75	11	267	331	203	139	395	459	1880
8	72	328	264	136	200	456	392	1856
509	445	189	253	381	317	61	125	2280
434	498	242	178	306	370	114	50	2192
2052	2052	2052	2052	2052	2052	2052	2052	

Table 8

295	39	103	487	423	167	231	359	2104
364	108	44	428	492	236	172	300	2144
145	401	465	81	17	273	337	209	1928
222	478	414	30	94	350	286	158	2032
5	261	325	197	133	389	453	69	1832
74	330	266	138	202	458	394	10	1872
435	179	243	371	307	51	115	499	2200
512	256	192	320	384	128	64	448	2304
2052	2052	2052	2052	2052	2052	2052	2052	

Fig 45 Z aspect

The subsets in this cube comprise 4x2 squares either by row or column which produce the constant.

	Constant Count			
	X aspect	Y aspect	Z aspect	Total
Rows	64	64		128
Columns	64		64	128
Fwd Diagonals	64		64	128
Bwd Diagonals	64	64	64	192
Horizontal Subsets	512			512
Vertical Subsets	512		128	640
Pillars		64	64	128
Cubic Diagonals				
Total	1,280	192	384	1,856

For a Magic Cube to have no cubic diagonals is an oxymoron too far, so let's move on to the second method of transforming the pandiagonal Magic Square into a Magic Cube.

I call it **the boustrophedon method**. Boustrophedon comes from the Greek and describes the method by which an ox would plough a field – from left to right and back again from right to left. We'll use the same Magic Square as before to create an entirely different and hopefully more successful Magic Cube.

1	14	55	60	35	48	21	26	260
15	4	57	54	45	34	27	24	260
58	53	16	3	28	23	46	33	260
56	59	2	13	22	25	36	47	260
18	29	40	43	52	63	6	9	260
32	19	42	37	62	49	12	7	260
41	38	31	20	11	8	61	50	260
39	44	17	30	5	10	51	64	260
260	260	260	260	260	260	260	260	

Fig 46

I use the square as a guide to position the numbers 1 to 512 in eight 8x8 grids. I begin by placing the numbers 1 to 8 in turn in the first cell of each grid, that is, the cell position of the number 1 in the Magic Square. Then, in boustrophedon fashion, and taking care to position the number 9 in the same cell position as the number 2 in the magic Square, I place the numbers 9 to 16 and return back to the first grid, each number maintaining the same cell position. The number 16 is now in the same cell position of number 2 in the magic square. The next number, 17, now takes the cell position of number 3 in the magic square and the sequence continues back to the eighth grid and number 24. The position of number 25 is determined by the cell position of number 4 in the magic square, and the process continues until we have the number 32 in the first grid. Let us now take stock. We have eight grids each containing four numbers adding up to 66 and occupying the cell positions of the numbers 1 to 4 of the Magic Square. Can you see what's going on? We are cloning the Magic Square!

The boustrophedon sequence continues with the number 33 starting from the cell position of number 5 in the Magic Square. When the sequence is completed we will have an 8x8 Magic Cube, each slice possessing all the qualities of the Magic Square. The boustrophedon method always achieves a perfect Magic Cube even though the pillars necessarily fail. However, it works only with even-numbered magic squares. The ox has always to return to whence it began in order to eat its oats. With an odd-numbered square it would remain stranded at the far end of the field.

Square 1

1	112	433	480	273	384	161	208	2052
113	32	449	432	353	272	209	192	2052
464	417	128	17	224	177	368	257	2052
448	465	16	97	176	193	288	369	2052
144	225	320	337	416	497	48	65	2052
256	145	336	289	496	385	96	49	2052
321	304	241	160	81	64	481	400	2052
305	352	129	240	33	80	401	512	2052
2052	2052	2052	2052	2052	2052	2052	2052	

Square 2

2	111	434	479	274	383	162	207	2052
114	31	450	431	354	271	210	191	2052
463	418	127	18	223	178	367	258	2052
447	466	15	98	175	194	287	370	2052
143	226	319	338	415	498	47	66	2052
255	146	335	290	495	386	95	50	2052
322	303	242	159	82	63	482	399	2052
306	351	130	239	34	79	402	511	2052
2052	2052	2052	2052	2052	2052	2052	2052	

Square 3

3	110	435	478	275	382	163	206	2052
115	30	451	430	355	270	211	190	2052
462	419	126	19	222	179	366	259	2052
446	467	14	99	174	195	286	371	2052
142	227	318	339	414	499	46	67	2052
254	147	334	291	494	387	94	51	2052
323	302	243	158	83	62	483	398	2052
307	350	131	238	35	78	403	510	2052
2052	2052	2052	2052	2052	2052	2052	2052	

Square 4

4	109	436	477	276	381	164	205	2052
116	29	452	429	356	269	212	189	2052
461	420	125	20	221	180	365	260	2052
445	468	13	100	173	196	285	372	2052
141	228	317	340	413	500	45	68	2052
253	148	333	292	493	388	93	52	2052
324	301	244	157	84	61	484	397	2052
308	349	132	237	36	77	404	509	2052
2052	2052	2052	2052	2052	2052	2052	2052	

Square 5

5	108	437	476	277	380	165	204	2052
117	28	453	428	357	268	213	188	2052
460	421	124	21	220	181	364	261	2052
444	469	12	101	172	197	284	373	2052
140	229	316	341	412	501	44	69	2052
252	149	332	293	492	389	92	53	2052
325	300	245	156	85	60	485	396	2052
309	348	133	236	37	76	405	508	2052
2052	2052	2052	2052	2052	2052	2052	2052	

Square 6

6	107	438	475	278	379	166	203	2052
118	27	454	427	358	267	214	187	2052
459	422	123	22	219	182	363	262	2052
443	470	11	102	171	198	283	374	2052
139	230	315	342	411	502	43	70	2052
251	150	331	294	491	390	91	54	2052
326	299	246	155	86	59	486	395	2052
310	347	134	235	38	75	406	507	2052
2052	2052	2052	2052	2052	2052	2052	2052	

Square 7

7	106	439	474	279	378	167	202	2052
119	26	455	426	359	266	215	186	2052
458	423	122	23	218	183	362	263	2052
442	471	10	103	170	199	282	375	2052
138	231	314	343	410	503	42	71	2052
250	151	330	295	490	391	90	55	2052
327	298	247	154	87	58	487	394	2052
311	346	135	234	39	74	407	506	2052
2052	2052	2052	2052	2052	2052	2052	2052	

Square 8

8	105	440	473	280	377	168	201	2052
120	25	456	425	360	265	216	185	2052
457	424	121	24	217	184	361	264	2052
441	472	9	104	169	200	281	376	2052
137	232	313	344	409	504	41	72	2052
249	152	329	296	489	392	89	56	2052
328	297	248	153	88	57	488	393	2052
312	345	136	233	40	73	408	505	2052
2052	2052	2052	2052	2052	2052	2052	2052	

Fig 47

All the squares are pandiagonal and together produce 64 cubic diagonals. **But that's not all!** An interesting feature is that the diagonals of the squares' quarters total either 1024 or 1028, but if we transpose columns 1 and four, and columns 5 and eight we make all those diagonals equal with incredible consequences.

112	433	480	1	384	161	208	273	2052
32	449	432	113	272	209	192	353	2052
417	128	17	464	177	368	257	224	2052
465	16	97	448	193	288	369	176	2052
225	320	337	144	497	48	65	416	2052
145	336	289	256	385	96	49	496	2052
304	241	160	321	64	481	400	81	2052
352	129	240	305	80	401	512	33	2052
2052	2052	2052	2052	2052	2052	2052	2052	

111	434	479	2	383	162	207	274	2052
31	450	431	114	271	210	191	354	2052
418	127	18	463	178	367	258	223	2052
466	15	98	447	194	287	370	175	2052
226	319	338	143	498	47	66	415	2052
146	335	290	255	386	95	50	495	2052
303	242	159	322	63	482	399	82	2052
351	130	239	306	79	402	511	34	2052
2052	2052	2052	2052	2052	2052	2052	2052	

110	435	478	3	382	163	206	275	2052
30	451	430	115	270	211	190	355	2052
419	126	19	462	179	366	259	222	2052
467	14	99	446	195	286	371	174	2052
227	318	339	142	499	46	67	414	2052
147	334	291	254	387	94	51	494	2052
302	243	158	323	62	483	398	83	2052
350	131	238	307	78	403	510	35	2052
2052	2052	2052	2052	2052	2052	2052	2052	

109	436	477	4	381	164	205	276	2052
29	452	429	116	269	212	189	356	2052
420	125	20	461	180	365	260	221	2052
468	13	100	445	196	285	372	173	2052
228	317	340	141	500	45	68	413	2052
148	333	292	253	388	93	52	493	2052
301	244	157	324	61	484	397	84	2052
349	132	237	308	77	404	509	36	2052
2052	2052	2052	2052	2052	2052	2052	2052	

108	437	476	5	380	165	204	277	2052
28	453	428	117	268	213	188	357	2052
421	124	21	460	181	364	261	220	2052
469	12	101	444	197	284	373	172	2052
229	316	341	140	501	44	69	412	2052
149	332	293	252	389	92	53	492	2052
300	245	156	325	60	485	396	85	2052
348	133	236	309	76	405	508	37	2052
2052	2052	2052	2052	2052	2052	2052	2052	

107	438	475	6	379	166	203	278	2052
27	454	427	118	267	214	187	358	2052
422	123	22	459	182	363	262	219	2052
470	11	102	443	198	283	374	171	2052
230	315	342	139	502	43	70	411	2052
150	331	294	251	390	91	54	491	2052
299	246	155	326	59	486	395	86	2052
347	134	235	310	75	406	507	38	2052
2052	2052	2052	2052	2052	2052	2052	2052	

106	439	474	7	378	167	202	279	2052
26	455	426	119	266	215	186	359	2052
423	122	23	458	183	362	263	218	2052
471	10	103	442	199	282	375	170	2052
231	314	343	138	503	42	71	410	2052
151	330	295	250	391	90	55	490	2052
298	247	154	327	58	487	394	87	2052
346	135	234	311	74	407	506	39	2052
2052	2052	2052	2052	2052	2052	2052	2052	

105	440	473	8	377	168	201	280	2052
25	456	425	120	265	216	185	360	2052
424	121	24	457	184	361	264	217	2052
472	9	104	441	200	281	376	169	2052
232	313	344	137	504	41	72	409	2052
152	329	296	249	392	89	56	489	2052
297	248	153	328	57	488	393	88	2052
345	136	233	312	73	408	505	40	2052
2052	2052	2052	2052	2052	2052	2052	2052	

Fig 48 X aspect

89

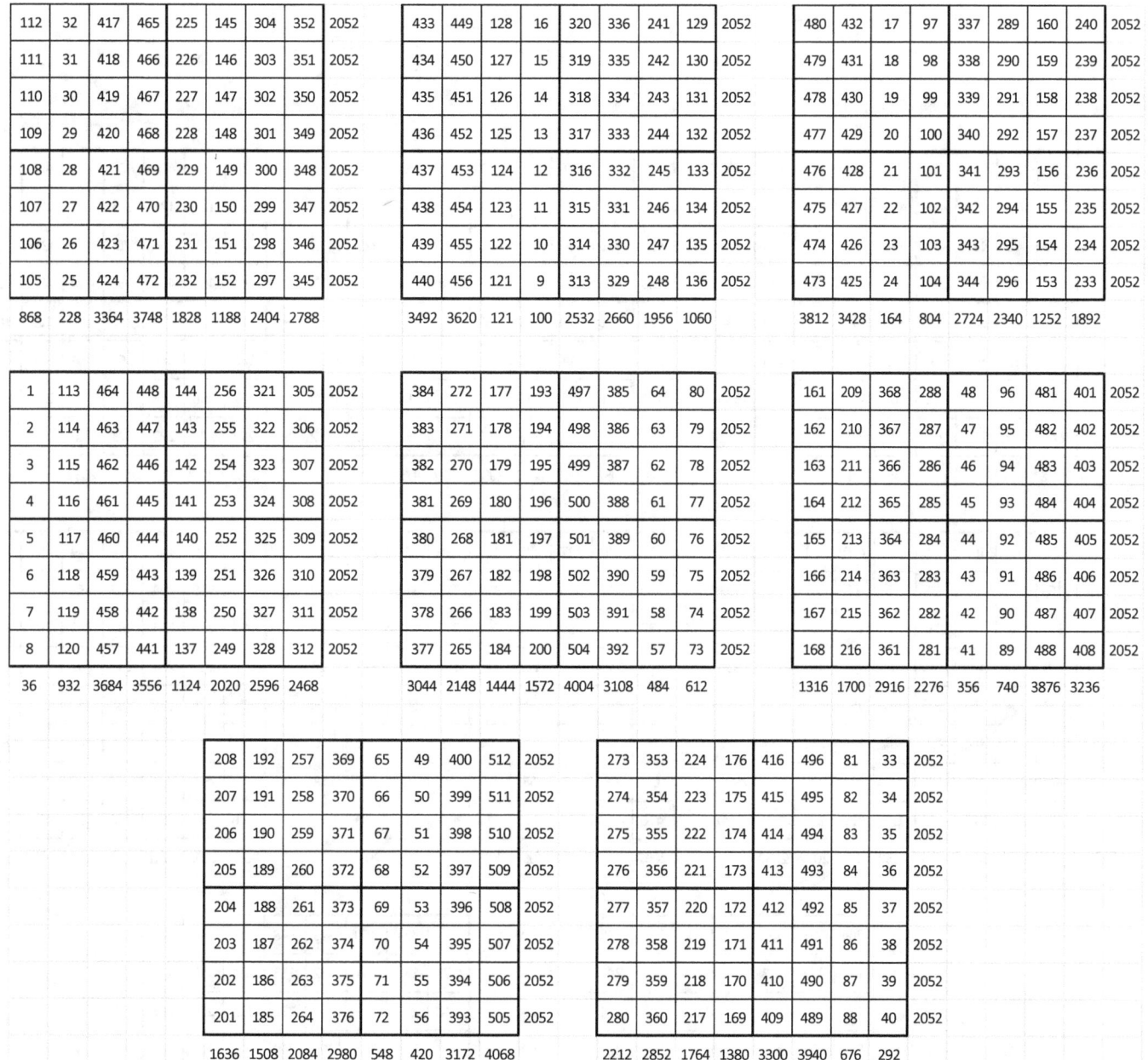

Fig 48 Y aspect

90

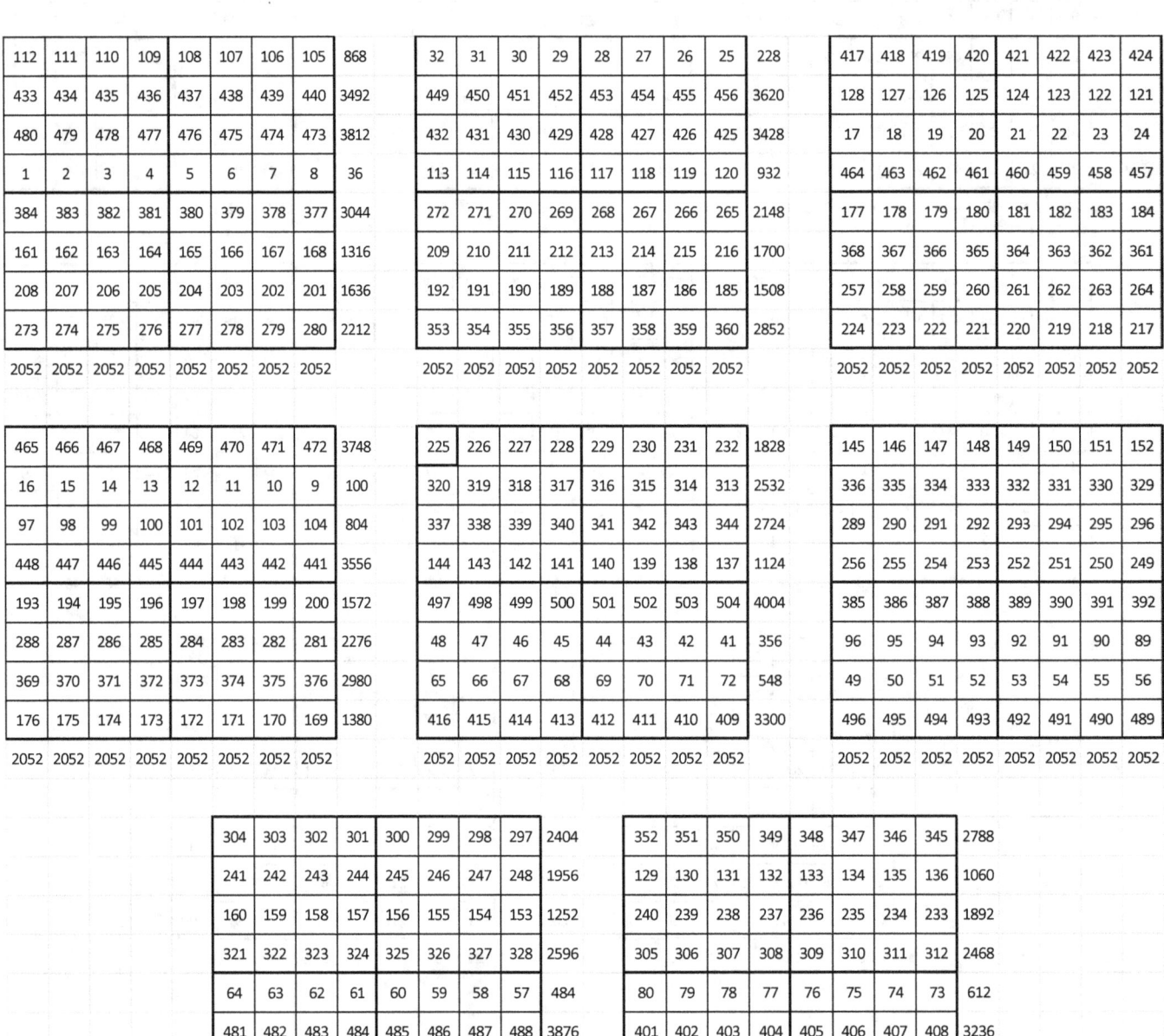

Fig 48 Z aspect

So how did this boustrophedon cube fare?

	X aspect	Y aspect	Z aspect	Total
. Constant Count				
Rows	64	64		128
Columns	64		64	128
Fwd Diagonals	64	64		128
Bwd Diagonals	64	64		128
Fwd Chevron Diagonals	512		512	1,024
Bwd Chevron Diagonals	512		512	1,024
Horizontal Subsets	512	128		640
Vertical Subsets	512		128	640
Pillars		64	64	128
Cubic Diagonals	32	32	32	96
Total	2,336	416	1,312	4,064

The half-constant of 1026 is produced in the quarters of all planar squares by their 4 rows, 4 columns and 2 diagonals. The cube thus comprises 32 arithmetically identical quarters with a constant of 1026 and so can be constructed in 32 x 32 ways, or 8192 ways if you count rotations and mirror images. This means they could be shuffled haphazardly in any way one chooses and would always end up as a magic cube, though whether the pillars and cubic diagonals would work would be a matter of luck. The permutations are so numerous that I won't bother to work out this cube's true score!

I am amazed that 512 numbers could be so arranged that there would be over four thousand discrete ways of uncovering their unique constant.

To the best of my knowledge and belief this is the first time an 8x8 Magic cube of this complexity has been published.

Before we encounter the chessboard puzzle which follows I cannot pass up the chance to retell the story of the servant who requested his ruler to reward his services by placing one grain of wheat on the first square of the chessboard, two on the second, four on the third, eight on the fourth and so on, doubling the number with each square up to the sixty fourth,. The ruler thought his servant was asking for a meagre reward until his treasurer advised him that it would bankrupt the realm.

The treasurer didn't know the half of it!

In total the number of grains that would have been needed is 18 followed by 18 noughts and would weigh one point two quadrillion metric tons, roughly 1,600 times the global wheat production today.

* * *

Puzzle Corner

You may be wondering what became of the young knight errant who was set the task of traversing a chequered courtyard in chess-knight fashion in order to free the King's daughter from Bandar's wicked spell. Well, I can tell you that his success and valour were rewarded by the young maiden's hand in marriage, but the evil Bandar could not forgive him for frustrating his fiendish fantasy, and when in due course the young couple's union was blessed with the birth of a son, Bandar's bitterness knew no bounds. Seizing upon an unguarded moment he plunged the babe into a stuporous slumber from which none could stir him.

"Well, young knight, you thwarter of schemes!" screamed Bandar, "Let me see you lift this spell! The stakes are raised and the task more testing, for now, entering upon the courtyard here and moving only as a chess-knight moves, you must number each square in turn so that every row, both east to west and north to south totals 260. Furthermore you must finish within a knight's move of the square from which you set forth, and from which I shall follow your frustration and failure with relish!"

"'Tis well that I should finish there," responded the brave knight, "for, by Our Lady, I can scarce forbear to cleave you when my ask is done." So saying, he crossed himself, drew his sword and stepped boldly into the courtyard.

To help you trace the young knight's steps the fourth of all his moves is given, together with the additional information that each quartet of moves occurred in the same quarter of the courtyard.

		49					
			13				
		33		17		21	
			61		45		9
1		37		53			
	29		25		41		
				5			
					57		

Puzzle No.11

9x9 Magic Cubes

I experienced many frustrations when I first attempted to create a 9x9 Magic Cube, similar to those I faced with the 9x9 Magic Square. The Latin square method proved a failure, as did the Successive Additive method. Eventually two thoughts occurred to me, so obvious were they that I felt cross with myself for having wasted time. I will deal with those thought processes in order, but the first needs some prior explanation.

I was once a member of a U3A group studying Ancient Greek History, Art and Literature, in the course of which we came across an enigmatic passage in Plato's "Republic" which stated among other things that a philosopher king is 729 times happier than a tyrant, and that one could verify that by working out the cube. Many suggestions had been made in the past about this obscure passage but no satisfactory solution has been found, so I thought it would be a good idea for me to try. The cube root of 729 is 9 so I constructed a 27x27 square containing nine 9x9 pandiagonal squares using the numbers 1 to 729 (**fig 49** below). The first square contained the numbers 1 to 81 and this was extrapolated to the other eight squares by sequentially adding 81. Each square had its own constant of, relatively, 369, 1098, 1827, 2556, 3285, 4014, 4743, 5472 and 6201. I studied the result and found a plausible link between each square and Plato's further statements and submitted an article to the editor of 'Omnibus' at St Hilda's College, Oxford, which published it in January, 2005. For those readers who may be interested I have re-produced the article at the end of this book.

49	5	71	47	7	69	54	3	64	130	86	152	128	88	150	135	84	145	211	167	233	209	169	231	216	165	226	3294
63	21	37	58	23	44	56	25	42	144	102	118	139	104	125	137	106	123	225	183	199	220	185	206	218	187	204	3294
29	79	15	36	75	10	31	77	17	110	160	96	117	156	91	112	158	98	191	241	177	198	237	172	193	239	179	3294
4	68	53	2	70	51	9	66	46	85	149	134	83	151	132	90	147	127	166	230	215	164	232	213	171	228	208	3294
27	39	55	22	41	62	20	43	60	108	120	136	103	122	143	101	124	141	189	201	217	184	203	224	182	205	222	3294
74	16	33	81	12	28	76	14	35	155	97	114	162	93	109	157	95	116	236	178	195	243	174	190	238	176	197	3294
67	50	8	65	52	6	72	48	1	148	131	89	146	133	87	153	129	82	229	212	170	227	214	168	234	210	163	3294
45	57	19	40	59	26	38	61	24	126	138	100	121	140	107	119	142	105	207	219	181	202	221	188	200	223	186	3294
11	34	78	18	30	73	13	32	80	92	115	159	99	111	154	94	113	161	173	196	240	180	192	235	175	194	242	3294
292	248	314	290	250	312	297	246	307	373	329	395	371	331	393	378	327	388	454	410	476	452	412	474	459	408	469	9855
306	264	280	301	266	287	299	268	285	387	345	361	382	347	368	380	349	366	468	426	442	463	428	449	461	430	447	9855
272	322	258	279	318	253	274	320	260	353	403	339	360	399	334	355	401	341	434	484	420	441	480	415	436	482	422	9855
247	311	296	245	313	294	252	309	289	328	392	377	326	394	375	333	390	370	409	473	458	407	475	456	414	471	451	9855
270	282	298	265	284	305	263	286	303	351	363	379	346	365	386	344	367	384	432	444	460	427	446	467	425	448	465	9855
317	259	276	324	255	271	319	257	278	398	340	357	405	336	352	400	338	359	479	421	438	486	417	433	481	419	440	9855
310	293	251	308	295	249	315	291	244	391	374	332	389	376	330	396	372	325	472	455	413	470	457	411	477	453	406	9855
288	300	262	283	302	269	281	304	267	369	381	343	364	383	350	362	385	348	450	462	424	445	464	431	443	466	429	9855
254	277	321	261	273	316	256	275	323	335	358	402	342	354	397	337	356	404	416	439	483	423	435	478	418	437	485	9855
535	491	557	533	493	555	540	489	550	616	572	638	614	574	636	621	570	631	697	653	719	695	655	717	702	651	712	16416
549	507	523	544	509	530	542	511	528	630	588	604	625	590	611	623	592	609	711	669	685	706	671	692	704	673	690	16416
515	565	501	522	561	496	517	563	503	596	646	582	603	642	577	598	644	584	677	727	663	684	723	658	679	725	665	16416
490	554	539	488	556	537	495	552	532	571	635	620	569	637	618	576	633	613	652	716	701	650	718	699	657	714	694	16416
513	525	541	508	527	548	506	529	546	594	606	622	589	608	629	587	610	627	675	687	703	670	689	710	668	691	708	16416
560	502	519	567	498	514	562	500	521	641	583	600	648	579	595	643	581	602	722	664	681	729	660	676	724	662	683	16416
553	536	494	551	538	492	558	534	487	634	617	575	632	619	573	639	615	568	715	698	656	713	700	654	720	696	649	16416
531	543	505	526	545	512	524	547	510	612	624	586	607	626	593	605	628	591	693	705	667	688	707	674	686	709	672	16416
497	520	564	504	516	559	499	518	566	578	601	645	585	597	640	580	599	647	659	682	726	666	678	721	661	680	728	16416
7668	7668	7668	7668	7668	7668	7668	7668	7668	9855	9855	9855	9855	9855	9855	9855	9855	9855	12042	12042	12042	12042	12042	12042	12042	12042	12042	

Fig 49

The 27x27 square is not a Magic Square, but the thought occurred to me that if it could be converted into one, then the nine squares would form the basis of a Platonic cube. The first step was to change the order of the nine squares within the larger square, so I chose the obvious first option of putting them in Lo Shu order. I admit that I was not at all surprised that it worked since Lo Shu had already shown me that it plays a major role in 9 power square matters.

130	86	152	128	88	150	135	84	145	697	653	719	695	655	717	702	651	712	292	248	314	290	250	312	297	246	307	9855
144	102	118	139	104	125	137	106	123	711	669	685	706	671	692	704	673	690	306	264	280	301	266	287	299	268	285	9855
110	160	96	117	156	91	112	158	98	677	727	663	684	723	658	679	725	665	272	322	258	279	318	253	274	320	260	9855
85	149	134	83	151	132	90	147	127	652	716	701	650	718	699	657	714	694	247	311	296	245	313	294	252	309	289	9855
108	120	136	103	122	143	101	124	141	675	687	703	670	689	710	668	691	708	270	282	298	265	284	305	263	286	303	9855
155	97	114	162	93	109	157	95	116	722	664	681	729	660	676	724	662	683	317	259	276	324	255	271	319	257	278	9855
148	131	89	146	133	87	153	129	82	715	698	656	713	700	654	720	696	649	310	293	251	308	295	249	315	291	244	9855
126	138	100	121	140	107	119	142	105	693	705	667	688	707	674	686	709	672	288	300	262	283	302	269	281	304	267	9855
92	115	159	99	111	154	94	113	161	659	682	726	666	678	721	661	680	728	254	277	321	261	273	316	256	275	323	9855
535	491	557	533	493	555	540	489	550	373	329	395	371	331	393	378	327	388	211	167	233	209	169	231	216	165	226	9855
549	507	523	544	509	530	542	511	528	387	345	361	382	347	368	380	349	366	225	183	199	220	185	206	218	187	204	9855
515	565	501	522	561	496	517	563	503	353	403	339	360	399	334	355	401	341	191	241	177	198	237	172	193	239	179	9855
490	554	539	488	556	537	495	552	532	328	392	377	326	394	375	333	390	370	166	230	215	164	232	213	171	228	208	9855
513	525	541	508	527	548	506	529	546	351	363	379	346	365	386	344	367	384	189	201	217	184	203	224	182	205	222	9855
560	502	519	567	498	514	562	500	521	398	340	357	405	336	352	400	338	359	236	178	195	243	174	190	238	176	197	9855
553	536	494	551	538	492	558	534	487	391	374	332	389	376	330	396	372	325	229	212	170	227	214	168	234	210	163	9855
531	543	505	526	545	512	524	547	510	369	381	343	364	383	350	362	385	348	207	219	181	202	221	188	200	223	186	9855
497	520	564	504	516	559	499	518	566	335	358	402	342	354	397	337	356	404	173	196	240	180	192	235	175	194	242	9855
454	410	476	452	412	474	459	408	469	49	5	71	47	7	69	54	3	64	616	572	638	614	574	636	621	570	631	9855
468	426	442	463	428	449	461	430	447	63	21	37	58	23	44	56	25	42	630	588	604	625	590	611	623	592	609	9855
434	484	420	441	480	415	436	482	422	29	79	15	36	75	10	31	77	17	596	646	582	603	642	577	598	644	584	9855
409	473	458	407	475	456	414	471	451	4	68	53	2	70	51	9	66	46	571	635	620	569	637	618	576	633	613	9855
432	444	460	427	446	467	425	448	465	27	39	55	22	41	62	20	43	60	594	606	622	589	608	629	587	610	627	9855
479	421	438	486	417	433	481	419	440	74	16	33	81	12	28	76	14	35	641	583	600	648	579	595	643	581	602	9855
472	455	413	470	457	411	477	453	406	67	50	8	65	52	6	72	48	1	634	617	575	632	619	573	639	615	568	9855
450	462	424	445	464	431	443	466	429	45	57	19	40	59	26	38	61	24	612	624	586	607	626	593	605	628	591	9855
416	439	483	423	435	478	418	437	485	11	34	78	18	30	73	13	32	80	578	601	645	585	597	640	580	599	647	9855
9855	9855	9855	9855	9855	9855	9855	9855	9855	9855	9855	9855	9855	9855	9855	9855	9855	9855	9855	9855	9855	9855	9855	9855	9855	9855	9855	

Fig 50

The next step was to redistribute the figures equally throughout the 27x27 square so that all 9x9 square possessed the same constant, each square necessarily having to contain an equal number of figures from the other 8 squares. After a couple of false starts I found a successful method by creating the first column of the proposed 27x27 square by choosing the first, third, fifth, seventh, ninth, second, fourth, sixth and eight figure in each square moving down one row successively. The second column was formed by beginning at the third square and choosing the third, fifth, seventh, ninth, second, fourth, sixth, eighth and first figure in each square moving down one row successively. The third column was formed by beginning at the fifth square and choosing the fifth, seventh, ninth, second, fourth, sixth, eighth, first and third figure in each square moving down one row successively and so on to the end of the completed 27x27 square. To my immense satisfaction I found that each of the 9 squares proved to have a constant of 3285 and the large square proved to be a pandiagonal Magic Square of the 27th order with a constant of 9855.

I had created my first 9x9 Magic Cube which I set out below.

98

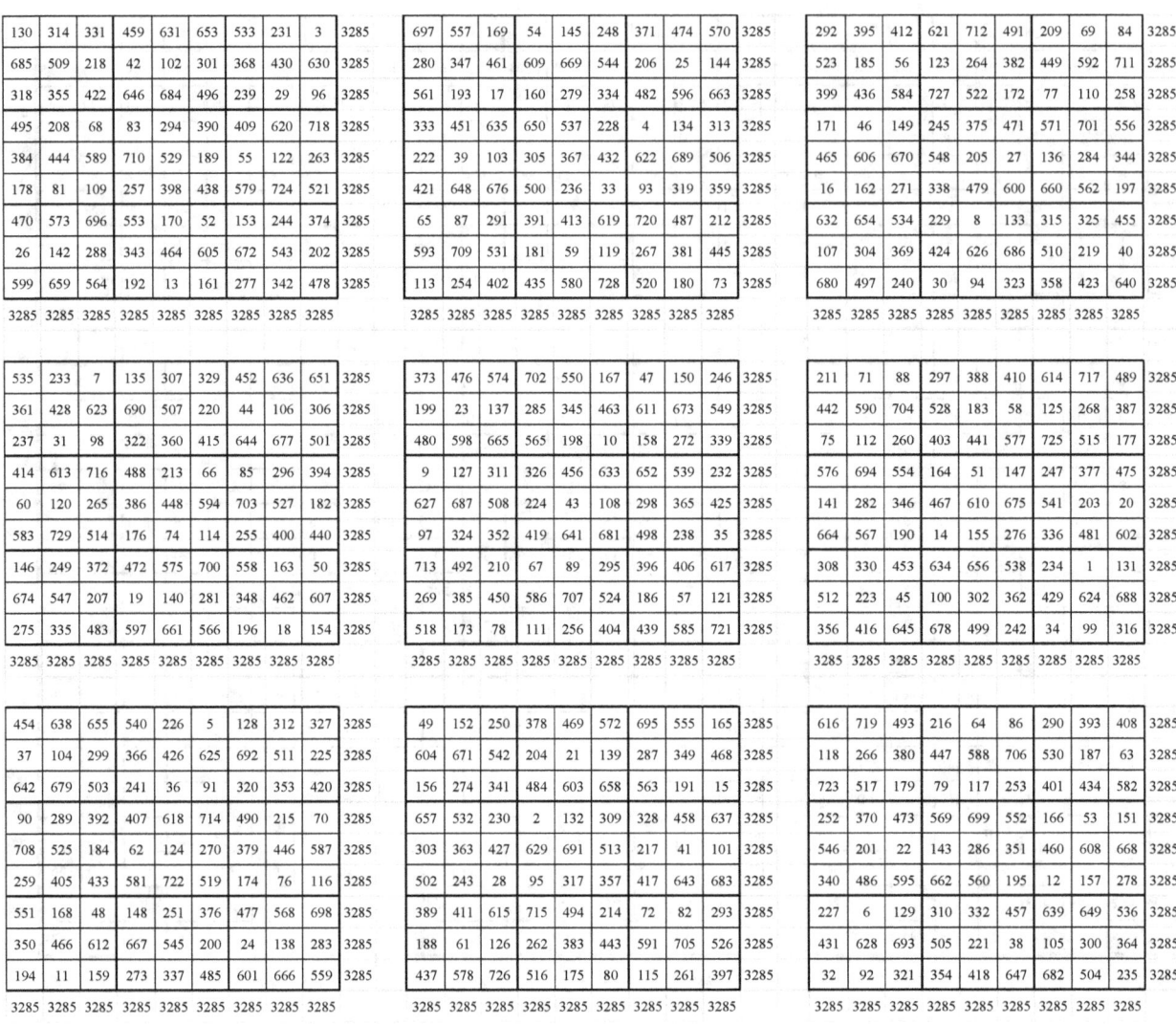

Fig 51 X aspect

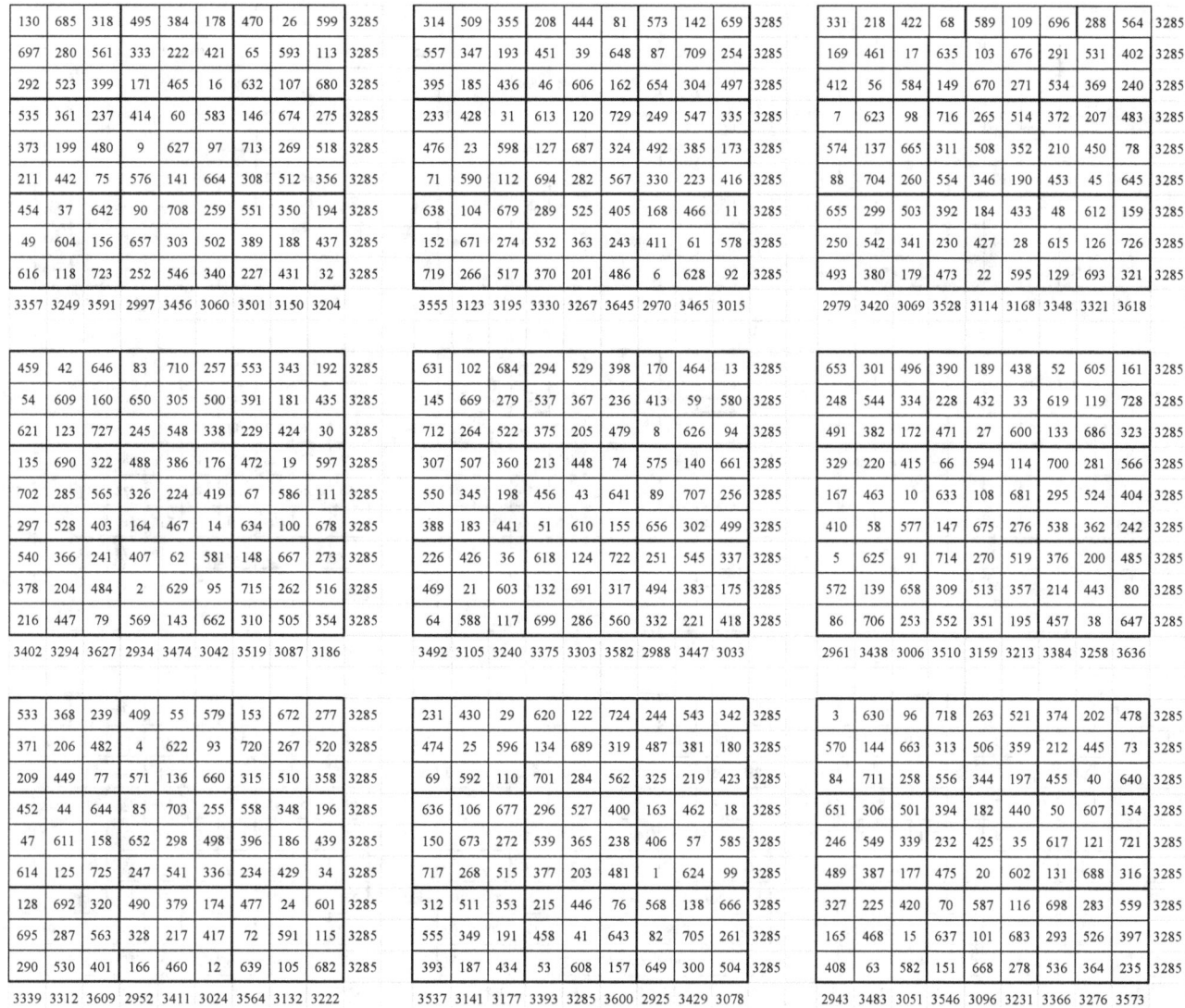

Fig 51　Y aspect

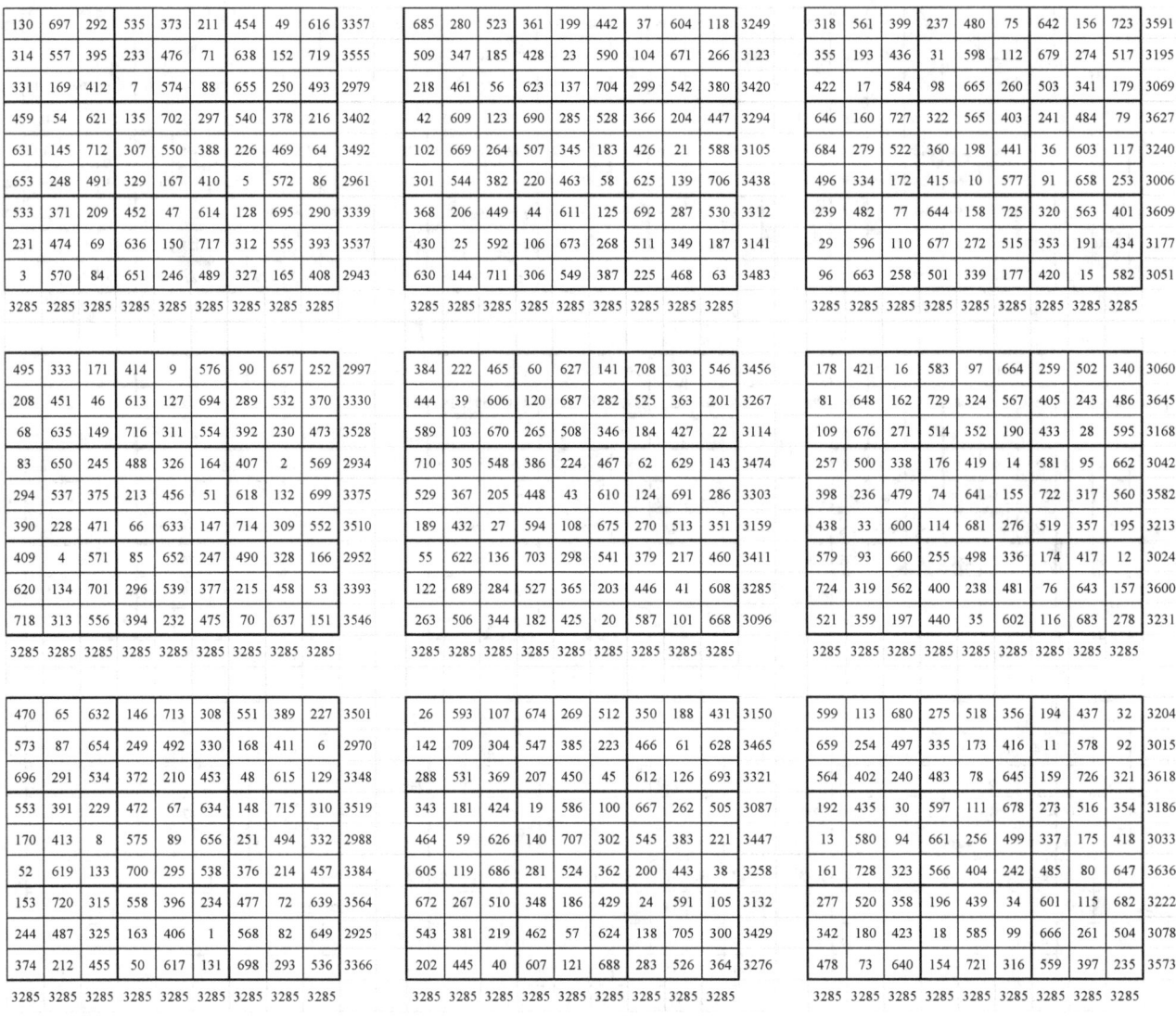

130	697	292	535	373	211	454	49	616	3357
314	557	395	233	476	71	638	152	719	3555
331	169	412	7	574	88	655	250	493	2979
459	54	621	135	702	297	540	378	216	3402
631	145	712	307	550	388	226	469	64	3492
653	248	491	329	167	410	5	572	86	2961
533	371	209	452	47	614	128	695	290	3339
231	474	69	636	150	717	312	555	393	3537
3	570	84	651	246	489	327	165	408	2943
3285	3285	3285	3285	3285	3285	3285	3285	3285	

685	280	523	361	199	442	37	604	118	3249
509	347	185	428	23	590	104	671	266	3123
218	461	56	623	137	704	299	542	380	3420
42	609	123	690	285	528	366	204	447	3294
102	669	264	507	345	183	426	21	588	3105
301	544	382	220	463	58	625	139	706	3438
368	206	449	44	611	125	692	287	530	3312
430	25	592	106	673	268	511	349	187	3141
630	144	711	306	549	387	225	468	63	3483
3285	3285	3285	3285	3285	3285	3285	3285	3285	

318	561	399	237	480	75	642	156	723	3591
355	193	436	31	598	112	679	274	517	3195
422	17	584	98	665	260	503	341	179	3069
646	160	727	322	565	403	241	484	79	3627
684	279	522	360	198	441	36	603	117	3240
496	334	172	415	10	577	91	658	253	3006
239	482	77	644	158	725	320	563	401	3609
29	596	110	677	272	515	353	191	434	3177
96	663	258	501	339	177	420	15	582	3051
3285	3285	3285	3285	3285	3285	3285	3285	3285	

495	333	171	414	9	576	90	657	252	2997
208	451	46	613	127	694	289	532	370	3330
68	635	149	716	311	554	392	230	473	3528
83	650	245	488	326	164	407	2	569	2934
294	537	375	213	456	51	618	132	699	3375
390	228	471	66	633	147	714	309	552	3510
409	4	571	85	652	247	490	328	166	2952
620	134	701	296	539	377	215	458	53	3393
718	313	556	394	232	475	70	637	151	3546
3285	3285	3285	3285	3285	3285	3285	3285	3285	

384	222	465	60	627	141	708	303	546	3456
444	39	606	120	687	282	525	363	201	3267
589	103	670	265	508	346	184	427	22	3114
710	305	548	386	224	467	62	629	143	3474
529	367	205	448	43	610	124	691	286	3303
189	432	27	594	108	675	270	513	351	3159
55	622	136	703	298	541	379	217	460	3411
122	689	284	527	365	203	446	41	608	3285
263	506	344	182	425	20	587	101	668	3096
3285	3285	3285	3285	3285	3285	3285	3285	3285	

178	421	16	583	97	664	259	502	340	3060
81	648	162	729	324	567	405	243	486	3645
109	676	271	514	352	190	433	28	595	3168
257	500	338	176	419	14	581	95	662	3042
398	236	479	74	641	155	722	317	560	3582
438	33	600	114	681	276	519	357	195	3213
579	93	660	255	498	336	174	417	12	3024
724	319	562	400	238	481	76	643	157	3600
521	359	197	440	35	602	116	683	278	3231
3285	3285	3285	3285	3285	3285	3285	3285	3285	

470	65	632	146	713	308	551	389	227	3501
573	87	654	249	492	330	168	411	6	2970
696	291	534	372	210	453	48	615	129	3348
553	391	229	472	67	634	148	715	310	3519
170	413	8	575	89	656	251	494	332	2988
52	619	133	700	295	538	376	214	457	3384
153	720	315	558	396	234	477	72	639	3564
244	487	325	163	406	1	568	82	649	2925
374	212	455	50	617	131	698	293	536	3366
3285	3285	3285	3285	3285	3285	3285	3285	3285	

26	593	107	674	269	512	350	188	431	3150
142	709	304	547	385	223	466	61	628	3465
288	531	369	207	450	45	612	126	693	3321
343	181	424	19	586	100	667	262	505	3087
464	59	626	140	707	302	545	383	221	3447
605	119	686	281	524	362	200	443	38	3258
672	267	510	348	186	429	24	591	105	3132
543	381	219	462	57	624	138	705	300	3429
202	445	40	607	121	688	283	526	364	3276
3285	3285	3285	3285	3285	3285	3285	3285	3285	

599	113	680	275	518	356	194	437	32	3204
659	254	497	335	173	416	11	578	92	3015
564	402	240	483	78	645	159	726	321	3618
192	435	30	597	111	678	273	516	354	3186
13	580	94	661	256	499	337	175	418	3033
161	728	323	566	404	242	485	80	647	3636
277	520	358	196	439	34	601	115	682	3222
342	180	423	18	585	99	666	261	504	3078
478	73	640	154	721	316	559	397	235	3573
3285	3285	3285	3285	3285	3285	3285	3285	3285	

Fig 51 Z aspect

Constant Count

	X aspect	Y aspect	Z aspect	Total
Rows	81	81		162
Columns	81		81	162
Fwd Diagonals	81		81	162
Bwd Diagonals	81	81	81	243
Subsets (3x3)	9	2	6	17
Pillars		81	81	162
Cubic Diagonals	81	81	81	243
	———	———	———	———
Total	414	326	411	1,151
	=====	=====	====	=======

To the best of my knowledge and belief this is the first time a 9x9 Magic cube of this complexity has been published.

My second obvious thought was that even though the Successive Additive method failed to produce a 9x9 Magic Cube, all 9x9 pandiagonal Magic Squares **must** consist of the Successive Additive couplings a0, b81, c162 and so on to i648. All I needed to do was to select any pandiagonal 9x9 Magic Square and analyse it with this purpose in mind. From my selection of 256 such squares I chose the following at random:

49	5	71	47	7	69	54	3	64	369
63	21	37	58	23	44	56	25	42	369
29	79	15	36	75	10	31	77	17	369
4	68	53	2	70	51	9	66	46	369
27	39	55	22	41	62	20	43	60	369
74	16	33	81	12	28	76	14	35	369
67	50	8	65	52	6	72	48	1	369
45	57	19	40	59	26	38	61	24	369
11	34	78	18	30	73	13	32	80	369
369	369	369	369	369	369	369	369	369	

Fig 52

Applying **a** to the numbers 1 to 9; **b** to the numbers 10 to 18; **c** to 19 to 27 and so on the Magic Square was revealed to be:

f	a	h	f	a	h	f	a	h
g	c	e	g	c	e	g	c	e
d	i	b	d	i	b	d	i	b
a	h	f	a	h	f	a	h	f
c	e	g	c	e	g	c	e	g
i	b	d	i	b	d	i	b	d
h	f	a	h	f	a	h	f	a
e	g	c	e	g	c	e	g	c
b	d	i	b	d	i	b	d	i

Fig 53

This enabled me to produce an amended additive chart, which written in figures is:

405	0	567	405	0	567	405	0	567	2916
486	162	324	486	162	324	486	162	324	2916
243	648	81	243	648	81	243	648	81	2916
0	567	405	0	567	405	0	567	405	2916
162	324	486	162	324	486	162	324	486	2916
648	81	243	648	81	243	648	81	243	2916
567	405	0	567	405	0	567	405	0	2916
324	486	162	324	486	162	324	486	162	2916
81	243	648	81	243	648	81	243	648	2916
2916	2916	2916	2916	2916	2916	2916	2916	2916	

Fig 54

Applying this to the Magic Square I produced the following 9x9 Magic Cube:

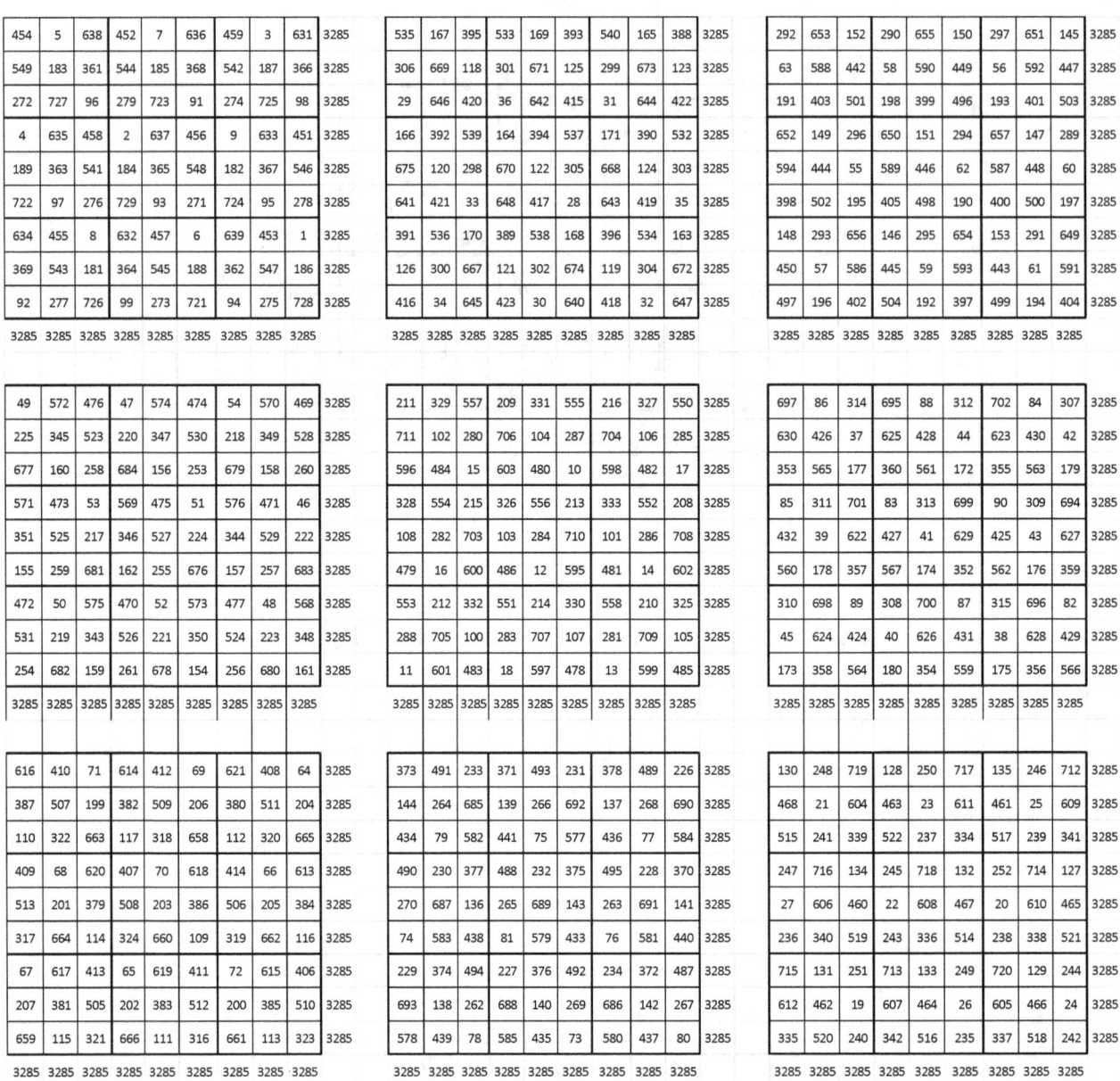

Fig 55 X aspect

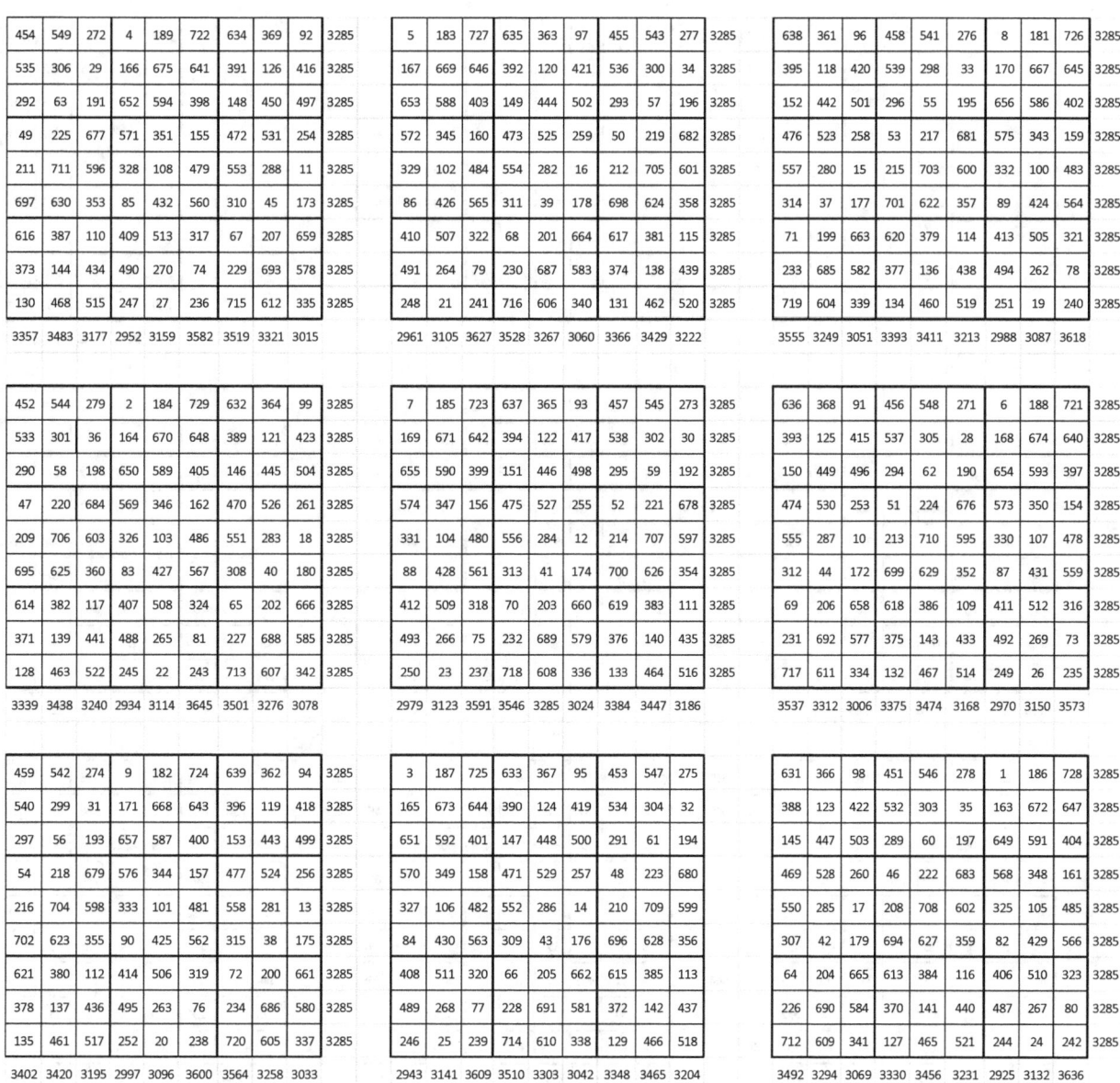

Fig 55 Y aspect

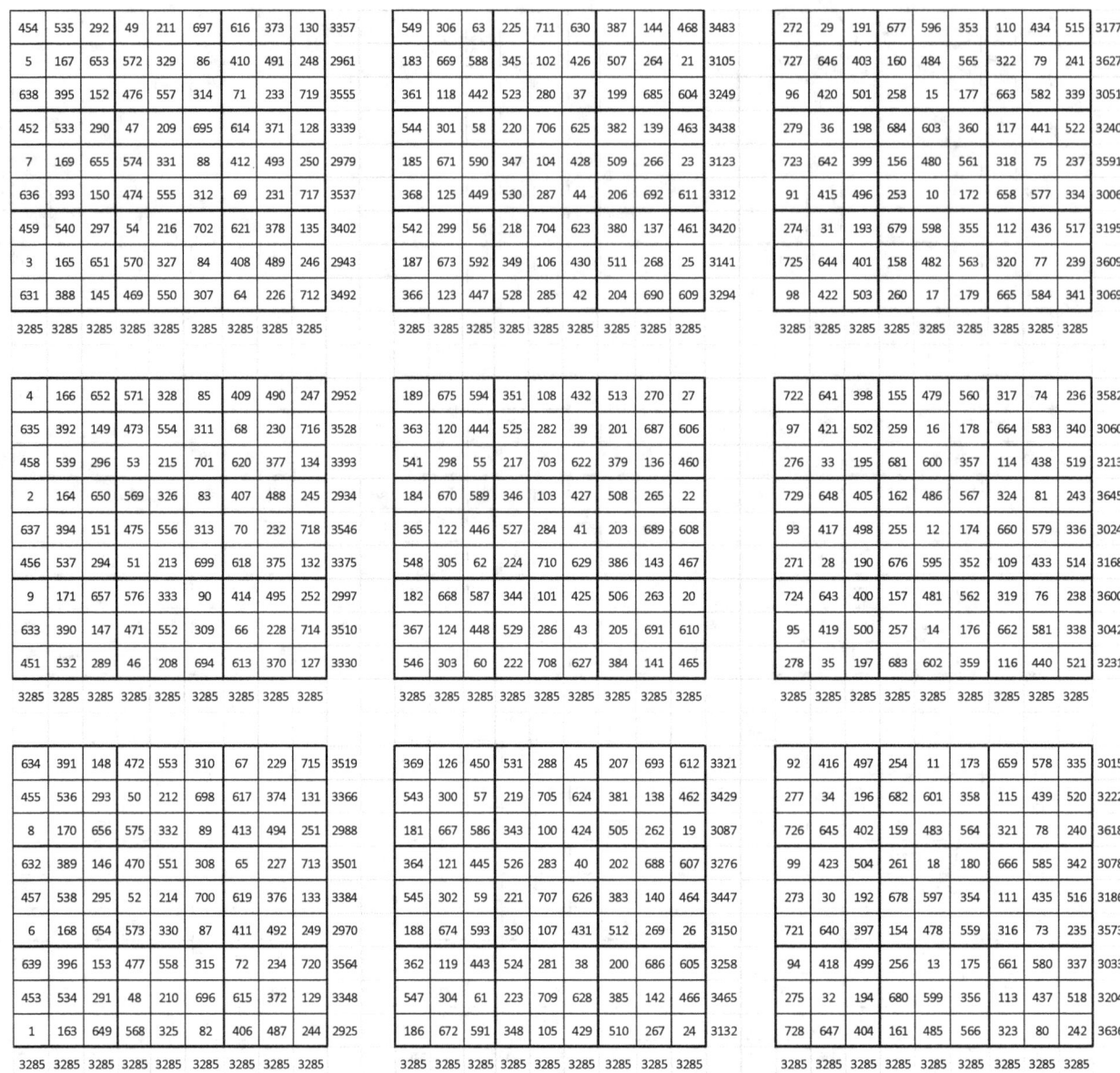

Fig 55 Z aspect

CONSTANT COUNT

	X aspect	Y aspect	Z aspect	Total
Rows	81	81		162
Columns	81		81	162
Fwd Diagonals	81			81
Bkwd Diagonals	81	81		162
Subsets (3x3)	81			81
Pillars		81	81	162
Cubic Diagonals	81	81	81	243
Total	486	324	243	1,053

To the best of my knowledge and belief this is the first time a 9x9 Magic cube of this complexity has been published.

Puzzle corner

You are required to supply the missing numbers from 1 to 81 so that the circumferences, radii and spirals give the constant of 369.

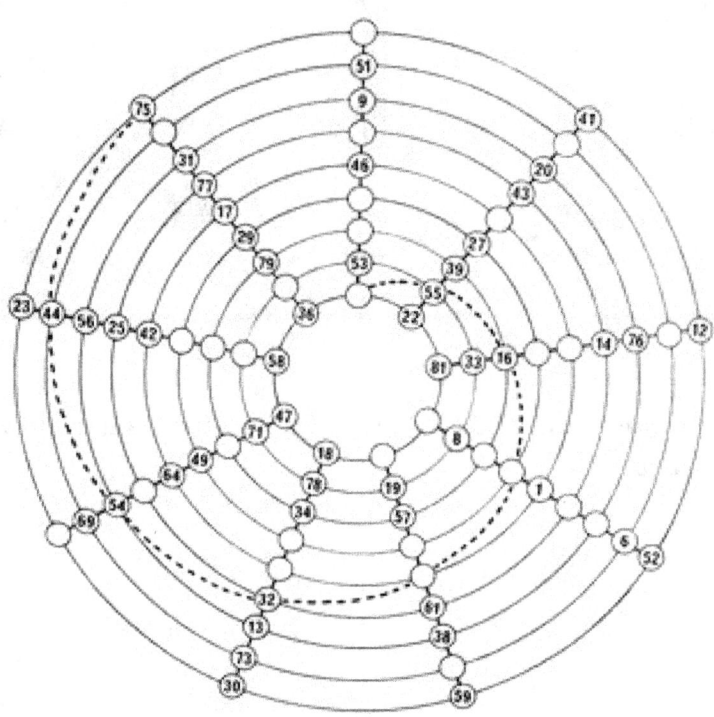

Puzzle No. 12

Solutions

1	15	10	8
12	6	3	13
7	9	16	2
14	4	5	11

Puzzle No. 1

8	20	3	22	12
10	25	7	21	2
9	11	13	15	17
24	5	19	1	16
14	4	23	6	18

Puzzle No. 2

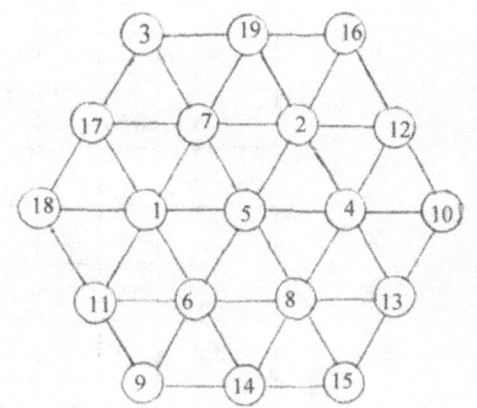

Puzzle No. 3

\

12	13	1	8
6	3	15	10
7	2	14	11
9	16	4	5

Puzzle No. 4

47	2	49	32	15	34	17	64
30	51	46	3	62	19	14	35
1	48	31	50	33	16	63	18
52	29	4	45	20	61	36	13
43	6	53	28	37	12	59	22
54	27	44	5	60	21	38	11
7	42	25	56	9	40	23	58
26	55	8	41	24	57	10	39

Puzzle No. 5

37	53	33	19	8	78	55	71	15
22	9	74	58	72	11	40	54	29
61	66	14	43	48	32	25	3	77
42	46	35	24	1	80	60	64	17
20	4	81	56	67	18	38	49	36
59	70	12	41	52	30	23	7	75
44	51	28	26	6	73	62	69	10
27	2	76	63	65	13	45	47	31
57	68	16	39	50	34	21	5	79

Puzzle No. 6

Puzzle No. 7

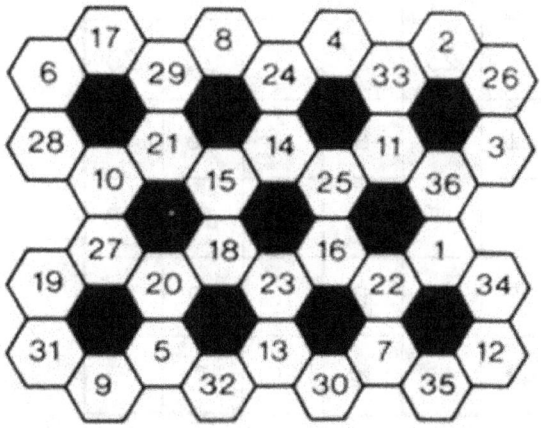

Puzzle No. 8

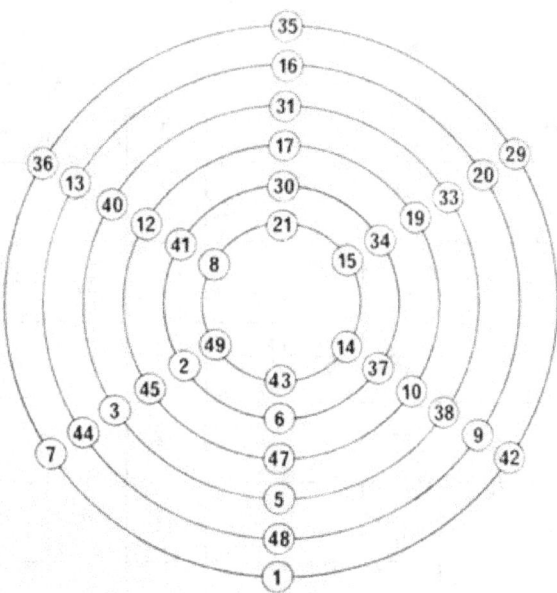

Puzzle No. 9

1	18	35	45	13	23	40
24	41	2	19	29	46	14
47	8	25	42	3	20	30
21	31	48	9	26	36	4
37	5	15	32	49	10	27
11	28	38	6	16	33	43
34	44	12	22	39	7	17

Puzzle No. 10

63	14	49	34	11	20	47	22
50	35	62	13	48	23	10	19
15	64	33	52	17	12	21	46
36	51	16	61	24	45	18	9
1	26	37	32	53	8	59	44
38	29	4	25	60	41	56	7
27	2	31	40	5	54	43	58
30	39	28	3	42	57	6	55

Puzzle No. 11

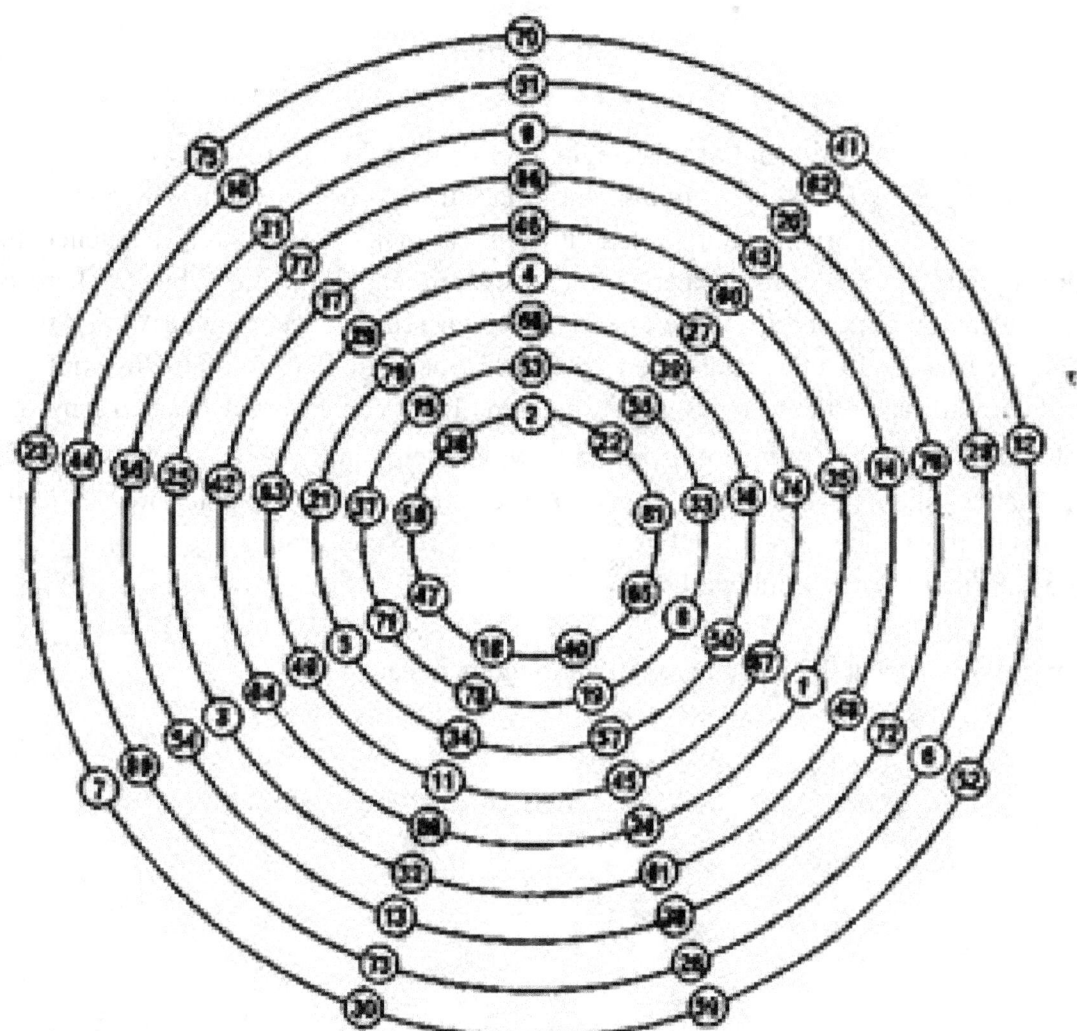

Puzzle No. 12

Summary

We have now reached the end of this Beginner's Guide which I trust you have enjoyed reading. On the other hand you may have found that it has cured your insomnia, which in itself is no bad thing. The most important point to bear in mind is that I have merely scratched the surface, writing mainly from my own experience and experiments with the subject. There is so much more that can be said, and for those interested to learn more I recommend W.S. Andrews' book, "Magic Squares and Cubes" published by Dover Books in 1917 and still in print one hundred years later, still as pertinent today as it was then. Time cannot upset the harmony, the balance, the logic and the stability of numbers, with their universal and enduring application. Mathematics, like its two brothers, Physics and Chemistry, is immutable throughout the cosmos. If a copy of this Guide were to find its way to a galactic schoolroom a gazillion light years away, all the Constant Counts would still apply.

Mmm. I wonder what the galactic royalty arrangements are.

* * * * *

A Magic Square for Plato

In Plato's "Republic" there is a celebrated passage pointing out the importance of the number 729. The passage (Book IX, §587-8 Lee's translation) reads:

"Do you know," I asked, "just how much unhappier the tyrant is than the philosopher king?"
"No, tell me."…………
"You will find, if you work out the cube, that the measure of difference between the two in terms of true pleasure is that the philosopher king lives seven hundred and twenty-nine times more pleasantly than the tyrant, and the tyrant the same amount more painfully than the philosopher king."
"What a terrific calculation," he exclaimed, "and all to show how much difference there is between the just and unjust man in terms of pleasure and pain!"
"But it's quite correct," I replied, "and fits human life, if human life is measured by days and nights and months and years."

Nowadays many would consider it irrational to ascribe attributes to numbers, and would hold that the number 729 has no more significance in human affairs than the number 42 which, in 'The Hitchhiker's Guide to the Galaxy' was given as the answer to Life, the Universe and Everything. However, in Plato's time mathematicians of immense stature, such as Pythagoras who said that number is the origin of all things, did indeed ascribe some ulterior significance, some almost supernatural or magical power to numbers. Plutarch set out the numbers attributed to the bodies in the solar system:- Antichthon, Earth's doppel-ganger, is 3; Earth itself is 9; the Moon 27; Mercury 81; Venus 243 and the Sun 729.

Plato's (or more likely Pythagoras') arithmetic is plain to see – he arrives at the figure of 729 by squaring the cube of three – but it is less clear how he saw this number fitting into human life. Many explanations have been offered in the past. One suggestion long ago was that a 27-power square, (i.e. one of 27 x 27 cells) with the numbers 1 to 729 inserted in numerical order would reveal the number 365 as the central square, clearly indicating the days in a year. From this it was suggested that the number 27 indicates the days in the lunar month, with the number 729 indicating the number of days and nights in a calendar year. It was further suggested that if the square were set out on a black and white chequerboard, the days and nights would be self-evident.

This suggested explanation, though attempting to answer the question of days, nights and months, fails to account either for the years, or for the reason for increased happiness. If indeed a 27-power square *was* an essential step in arriving at Plato's conclusion then it must have been a more complex square than one set out in numerical sequence. Could it have been a magic square? The evidence would appear to be against it. The earliest written evidence of Magic Squares appears in Chinese literature, written about 1100 AD, though it is known that Arabian astrologers were using them to prepare horoscopes two or three centuries earlier. Although there is no evidence of Magic Squares in use in nnAncient Greece, absence of evidence does not mean evidence of absence. It would have required no intellectual leap for a member of Pythagoras' school to configure the basic 3-power magic square, where the rows, columns and diagonals each total fifteen, as in fig 1.

2	9	4
7	5	3
6	1	8

fig 1

It is hardly likely that a school engrossed with numbers would have failed to configure such an elementary arrangement of numbers either by design or by chance. Even so, it would still have required an extraordinary extrapolation of the basic square to produce a 27-power square with the necessary properties to justify Plato's statement.

Plato implied that to find the solution one should work out the cube. The cube root of 729 is 9, so it seems probable that he was implying that the key to the solution is the number 9. A 27-power square contains nine 9-power squares within its borders, and each of these squares, if they are part of the key, must have a significance beyond being part of the larger square. Did Pythagoras' school devise such a complex square, and if so, could it have been similar to the one I have constructed in fig 2?

49	5	71	47	7	69	54	3	64	130	86	152	128	88	150	135	84	145	211	167	233	209	169	231	216	165	226	3294
63	21	37	58	23	44	56	25	42	144	102	118	139	104	125	137	106	123	225	183	199	220	185	206	218	187	204	3294
29	79	15	36	75	10	31	77	17	110	160	96	117	156	91	112	158	98	191	241	177	198	237	172	193	239	179	3294
4	68	53	2	70	51	9	66	46	85	149	134	83	151	132	90	147	127	166	230	215	164	232	213	171	228	208	3294
27	39	55	22	41	62	20	43	60	108	120	136	103	122	143	101	124	141	189	201	217	184	203	224	182	205	222	3294
74	16	33	81	12	28	76	14	35	155	97	114	162	93	109	157	95	116	236	178	195	243	174	190	238	176	197	3294
67	50	8	65	52	6	72	48	1	148	131	89	146	133	87	153	129	82	229	212	170	227	214	168	234	210	163	3294
45	57	19	40	59	26	38	61	24	126	138	100	121	140	107	119	142	105	207	219	181	202	221	188	200	223	186	3294
11	34	78	18	30	73	13	32	80	92	115	159	99	111	154	94	113	161	173	196	240	180	192	235	175	194	242	3294
292	248	314	290	250	312	297	246	307	373	329	395	371	331	393	378	327	388	454	410	476	452	412	474	459	408	469	9855
306	264	280	301	266	287	299	268	285	387	345	361	382	347	368	380	349	366	468	426	442	463	428	449	461	430	447	9855
272	322	258	279	318	253	274	320	260	353	403	339	360	399	334	355	401	341	434	484	420	441	480	415	436	482	422	9855
247	311	296	245	313	294	252	309	289	328	392	377	326	394	375	333	390	370	409	473	458	407	475	456	414	471	451	9855
270	282	298	265	284	305	263	286	303	351	363	379	346	365	386	344	367	384	432	444	460	427	446	467	425	448	465	9855
317	259	276	324	255	271	319	257	278	398	340	357	405	336	352	400	338	359	479	421	438	486	417	433	481	419	440	9855
310	293	251	308	295	249	315	291	244	391	374	332	389	376	330	396	372	325	472	455	413	470	457	411	477	453	406	9855
288	300	262	283	302	269	281	304	267	369	381	343	364	383	350	362	385	348	450	462	424	445	464	431	443	466	429	9855
254	277	321	261	273	316	256	275	323	335	358	402	342	354	397	337	356	404	416	439	483	423	435	478	418	437	485	9855
535	491	557	533	493	555	540	489	550	616	572	638	614	574	636	621	570	631	697	653	719	695	655	717	702	651	712	16416
549	507	523	544	509	530	542	511	528	630	588	604	625	590	611	623	592	609	711	669	685	706	671	692	704	673	690	16416
515	565	501	522	561	496	517	563	503	596	646	582	603	642	577	598	644	584	677	727	663	684	723	658	679	725	665	16416
490	554	539	488	556	537	495	552	532	571	635	620	569	637	618	576	633	613	652	716	701	650	718	699	657	714	694	16416
513	525	541	508	527	548	506	529	546	594	606	622	589	608	629	587	610	627	675	687	703	670	689	710	668	691	708	16416
560	502	519	567	498	514	562	500	521	641	583	600	648	579	595	643	581	602	722	664	681	729	660	676	724	662	683	16416
553	536	494	551	538	492	558	534	487	634	617	575	632	619	573	639	615	568	715	698	656	713	700	654	720	696	649	16416
531	543	505	526	545	512	524	547	510	612	624	586	607	626	593	605	628	591	693	705	667	688	707	674	686	709	672	16416
497	520	564	504	516	559	499	518	566	578	601	645	585	597	640	580	599	647	659	682	726	666	678	721	661	680	728	16416
7668	7668	7668	7668	7668	7668	7668	7668	7668	9855	9855	9855	9855	9855	9855	9855	9855	9855	12042	12042	12042	12042	12042	12042	12042	12042	12042	

fig 2

This square contains the numbers 1 to 729 in non-numerical order, with 365 occupying the centre square. It is not a Magic Square in itself but consists of nine 9-power Magic Squares, one containing the numbers 1 to 81, and the others with increments of 81 in succession.

Each of these smaller squares has a numerical significance. The rows, columns and diagonals in each individual square produce a total, or constant, as follows:

Square 1	369	=	1 year
2	1098	=	3 years
3	1827	=	5 years
4	2556	=	7 years
5	3285	=	9 years
6	4014	=	11 years
7	4743	=	13 years
8	5472	=	15 years
9	6201	=	17 years
Totals	29565		81 years

Each of the smaller squares within the large square thus indicates a certain age or period of time, and various combinations of these squares can indicate any period or age up to 81 years, a reasonable life span for an individual. These periods can themselves be set to form a 3-power square, or, as shown in fig 3, a fylfot.

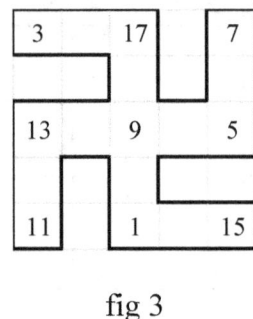

fig 3

The rows, columns and diagonals produce the constant 27, and the fylfot, a symbol of well-being and happiness, common in Ancient Greek designs, totals 81.

From all the above one can derive the following facts, which to modern eyes are merely the product of simple arithmetic and the re-arrangement of the various powers of 3, but to Ancient Greek eyes could well have had a magical or supernatural significance. It would have been seen by those eyes that:

the centre square of fig.2 is occupied by the number 365, indicating the days in a year;
the number of days and nights is indicated by the number 729;
the 27 columns and rows show the number of days in a lunar month;
the constant of each of the smaller squares, either alone or in combination, can be developed to indicate a certain number of years up to and including 81;
the constant for each smaller square is produced by totaling not only the columns, rows and diagonals but also any 3-power square within its borders (i.e. any nine numbers which together form a square);
the constant of the square in figure 3 is 27;

the total of the numbers in the fylfot, the symbol associated with well-being and happiness, is 81; this 81 is based upon a 3-power square, so in a 27-power square it must be 81 x 9 which equals 729.

Thus, with the aid of creative exegesis, it could be suggested that fig. 2 fulfils Plato's requirements of fitting human life as measured by days, nights, months and years, and the 729 times pleasure enjoyed by the philosopher king.

There is an even more enigmatic paragraph in Plato's "Republic" (Book VIII § 546 Lee's translation) in which he sets out the right and wrong times for breeding. It reads:

And though the Rulers you have trained for your city are wise, reason and observation will not always enable them to hit on the right and wrong times for breeding; some time they will miss them and then children will be begotten amiss. For the divine creature there is a period contained in a perfect number. For the human creature it is the smallest number in which certain multiplications, dominating and dominated, comprising three distances and four terms, give a final result, by making like and unlike, increasing and decreasing, which is commensurate and rational. Their basic ratio of four to three, coupled with five, and multiplied by three, yields two harmonies, of which one is the product of equal factors and of a hundred multiplied the same number of times, while the other is the product of factors of which some are equal, some unequal, that is, **either** *a hundred squares of diagonal of rational number, each diminished by one,* **or** *a hundred squares of irrational number, each diminished by two,* **and** *one hundred cubes of three.*

No satisfactory solution has been given for this obscure paragraph, though many suggestions have been made in the past. Whatever the calculation, however, it is clear that the answer, however produced, is intended to indicate the period(s) within the life span of a man and/or a woman when it would be most appropriate for the begetting of children. It is no more than idle speculation, but if a magic square similar to that in fig 2 *were* in being in ancient Greece, then, since it embraces periods, it could have provided a matrix for just such a calculation.

For those interested in Magic Squares, the nine smaller squares are pandiagonal. Any number of rows can be moved from top to bottom, and/or any number of columns can be moved from left to right, without destroying the constant produced by the diagonals.

Kenneth Kelsey

* * *

APPENDIX I

Diagonals of 5x5 cube Fig 31

24	91	38	110	52	315
37	109	51	23	95	315
55	109	38	92	21	315
93	40	107	54	21	315
106	22	38	54	95	315

28	125	67	14	81	315
66	13	85	27	124	315
84	26	123	70	12	315
122	69	11	83	30	315
15	82	29	121	68	315

57	125	38	76	19	315
100	13	51	119	32	315
113	26	94	7	75	315
1	69	107	50	88	315
44	82	25	63	101	315

86	33	105	72	19	315
104	71	18	90	32	315
17	89	31	103	75	315
35	102	74	16	88	315
73	20	87	34	101	315

120	4	38	72	81	315
8	42	51	90	124	315
46	60	94	103	12	315
64	98	107	16	30	315
77	111	25	34	68	315

24	91	38	110	52	315
28	125	67	14	81	315
57	4	96	43	115	315
86	33	105	72	19	315
120	62	9	76	48	315

49	116	63	10	77	315
53	25	92	39	106	315
82	29	121	68	15	315
111	58	5	97	44	315
20	87	34	101	73	315

74	116	38	85	2	315
78	25	67	114	31	315
107	29	96	18	65	315
11	58	105	47	94	315
45	87	9	51	123	315

99	41	113	60	2	315
103	75	17	89	31	315
7	79	46	118	65	315
36	108	55	22	94	315
70	12	84	26	123	315

124	66	13	85	27	315
3	100	42	114	56	315
32	104	71	18	90	315
61	8	80	47	119	315
95	37	109	51	23	315

24	91	38	110	52	315
49	116	63	10	77	315
74	16	88	35	102	315
99	41	113	60	2	315
124	66	13	85	27	315

37	109	51	23	95	315
62	9	76	48	120	315
87	34	101	73	20	315
112	59	1	98	45	315
12	84	26	123	70	315

55	22	94	36	108	315
80	47	119	61	8	315
105	72	19	86	33	315
5	97	44	111	58	315
30	122	69	11	83	315

93	40	107	54	21	315
118	65	7	79	46	315
18	90	32	104	71	315
43	115	57	4	96	315
68	15	82	29	121	315

106	53	25	92	39	315
6	78	50	117	64	315
31	103	75	17	89	315
56	3	100	42	114	315
81	28	125	67	14	315

Unbroken diagonals listed separately:

24	91	38	110	52	315
106	22	38	54	95	315
57	125	38	76	19	315
44	26	38	50	32	190
120	4	38	72	81	315
77	60	38	16	124	315
55	109	38	92	21	315
46	42	38	34	30	190

X aspect

24	91	38	110	52	315
120	62	9	76	48	315
74	116	38	85	2	315
45	29	38	47	31	190
124	16	38	60	77	315
95	54	38	22	106	315
57	125	38	76	19	315
32	50	38	26	44	190

Y aspect

24	91	38	110	52	315
124	16	38	60	77	315
55	109	38	92	21	315
30	34	38	42	46	190
106	22	38	54	95	315
81	72	38	4	120	315
74	116	38	85	2	315
31	47	38	29	45	190

Z aspect

APPENDIX II

Diagonals of 5x5 cube Fig 34

24	41	63	85	102	315
62	84	101	23	45	315
105	84	63	42	21	315
43	65	82	104	21	315
81	22	63	104	45	315

28	75	92	114	6	315
91	113	10	27	74	315
9	26	73	95	112	315
72	94	111	8	30	315
115	7	29	71	93	315

57	75	63	51	69	315
125	17	39	56	78	315
38	60	77	124	16	315
76	123	20	37	59	315
19	26	63	100	107	315

86	108	5	47	69	315
4	46	68	90	107	315
67	89	106	3	50	315
110	2	49	66	88	315
48	70	87	109	1	315

120	79	63	47	6	315
33	55	97	119	11	315
96	17	63	109	30	315
14	31	53	100	117	315
52	60	63	66	74	315

24	41	63	85	102	315
28	75	92	114	6	315
57	75	63	51	69	315
86	108	5	47	69	315
120	79	63	47	6	315

99	116	13	35	52	315
103	25	42	64	81	315
7	29	71	93	115	315
36	58	80	122	19	315
70	87	109	1	48	315

49	116	63	10	77	315
53	100	117	14	31	315
82	104	21	43	65	315
111	8	30	72	94	315
20	29	63	97	106	315

124	16	38	60	77	315
3	50	67	89	106	315
32	54	96	118	15	315
61	83	105	22	44	315
95	112	9	26	73	315

74	66	63	60	52	315
78	125	17	39	56	315
107	100	63	26	19	315
11	33	55	97	119	315
45	104	63	22	81	315

24	41	63	85	102	315
99	116	13	35	52	315
49	116	63	10	77	315
124	16	38	60	77	315
74	66	63	60	52	315

62	84	101	23	45	315
12	34	51	98	120	315
7	109	1	48	70	235
37	59	76	123	20	315
112	9	26	73	95	315

105	84	63	42	21	315
55	97	119	11	33	315
5	47	69	86	108	315
80	122	19	36	58	315
30	109	63	17	96	315

43	65	82	104	21	315
118	15	32	54	96	315
68	90	107	4	46	315
18	40	57	79	121	315
93	115	7	29	71	315

81	22	63	104	45	315
31	53	100	117	14	315
106	97	63	29	20	315
56	78	125	17	39	315
6	47	63	79	120	315

Unbroken diagonals listed separately:

24	41	63	85	102	315
81	22	63	104	45	315
57	75	63	51	69	315
52	60	63	66	74	315
120	79	63	47	6	315
19	26	63	100	107	315
105	84	63	42	21	315
96	17	63	109	30	315

X aspect

24	41	63	85	102	315
120	79	63	47	6	315
49	116	63	10	77	315
45	104	63	22	81	315
74	66	63	60	52	315
20	29	63	97	106	315
57	75	63	51	69	315
107	100	63	26	19	315

Y aspect

24	41	63	85	102	315
74	66	63	60	52	315
105	84	63	42	21	315
6	47	63	79	120	315
81	22	63	104	45	315
30	109	63	17	96	315
49	116	63	10	77	315
106	97	63	29	20	315

Z aspect

APPENDIX III

Diagonals of 7x7 cube Fig 41

1	264	128	342	157	77	235	1204
165	78	243	9	272	136	301	1204
280	144	302	173	86	202	17	1204
94	210	25	281	103	310	181	1204
111	318	189	53	211	33	289	1204
219	41	248	119	326	190	61	1204
334	149	69	227	49	256	120	1204
1204	1204	1204	1204	1204	1204	1204	

220	42	249	113	327	191	62	1204
335	150	70	228	43	257	121	1204
2	265	129	343	158	71	236	1204
166	79	244	10	273	137	295	1204
274	145	303	174	87	203	18	1204
95	204	26	282	104	311	182	1204
112	319	183	54	212	34	290	1204
1204	1204	1204	1204	1204	1204	1204	

96	205	27	283	105	312	176	1204
106	320	184	55	213	35	291	1204
221	36	250	114	328	192	63	1204
336	151	64	229	44	258	122	1204
3	266	130	337	159	72	237	1204
167	80	245	11	267	138	296	1204
275	146	304	175	88	197	19	1204
1204	1204	1204	1204	1204	1204	1204	

168	81	239	12	268	139	297	1204
276	147	305	169	89	198	20	1204
97	206	28	284	99	313	177	1204
107	321	185	56	214	29	292	1204
222	37	251	115	329	193	57	1204
330	152	65	230	45	259	123	1204
4	260	131	338	160	73	238	1204
1204	1204	1204	1204	1204	1204	1204	

331	153	66	231	46	253	124	1204
5	261	132	339	161	74	232	1204
162	82	240	13	269	140	298	1204
277	141	306	170	90	199	21	1204
98	207	22	285	100	314	178	1204
108	322	186	50	215	30	293	1204
223	38	252	116	323	194	58	1204
1204	1204	1204	1204	1204	1204	1204	

109	316	187	51	216	31	294	1204
224	39	246	117	324	195	59	1204
332	154	67	225	47	254	125	1204
6	262	133	340	155	75	233	1204
163	83	241	14	270	134	299	1204
278	142	307	171	91	200	15	1204
92	208	23	286	101	315	179	1204
1204	1204	1204	1204	1204	1204	1204	

279	143	308	172	85	201	16	1204
93	209	24	287	102	309	180	1204
110	317	188	52	217	32	288	1204
218	40	247	118	325	196	60	1204
333	148	68	226	48	255	126	1204
7	263	127	341	156	76	234	1204
164	84	242	8	271	135	300	1204
1204	1204	1204	1204	1204	1204	1204	

X aspect

1	264	128	342	157	77	235	1204
220	42	249	113	327	191	62	1204
96	5	27	283	105	312	176	1004
168	81	239	12	268	139	297	1204
331	153	66	231	46	253	124	1204
109	316	187	51	216	31	294	1204
279	143	308	172	85	201	16	1204
1204	1004	1204	1204	1204	1204	1204	

84	242	8	271	135	300	164	1204
149	69	227	49	256	120	334	1204
319	183	54	212	34	290	112	1204
146	304	175	88	197	19	275	1204
260	131	338	160	73	238	4	1204
38	252	116	323	194	58	223	1204
208	23	286	101	315	179	92	1204
1204	1204	1204	1204	1204	1204	1204	

307	171	91	200	15	278	142	1204
127	341	156	76	234	7	263	1204
248	119	326	190	61	219	41	1204
26	282	104	311	182	95	204	1204
245	11	267	138	296	167	80	1204
65	230	45	259	123	330	152	1204
186	50	215	30	293	108	322	1204
1204	1204	1204	1204	1204	1204	1204	

285	100	314	178	98	207	22	1204
14	270	134	299	163	83	241	1204
226	48	255	126	333	148	68	1204
53	211	33	289	111	318	189	1204
174	87	203	18	274	145	303	1204
337	159	72	237	3	266	130	1204
115	329	193	57	222	37	251	1204
1204	1204	1204	1204	1204	1204	1204	

214	29	292	107	321	185	56	1204
90	199	21	277	141	306	170	1204
155	75	233	6	262	133	340	1204
325	196	60	218	40	247	118	1204
103	310	181	94	210	25	281	1204
273	137	295	166	79	244	10	1204
44	258	122	336	151	64	229	1204
1204	1204	1204	1204	1204	1204	1204	

192	63	221	36	250	114	328	1204
313	177	97	206	28	284	99	1204
143	298	162	82	240	13	269	1207
254	125	332	154	67	225	47	1204
32	288	110	317	188	52	217	1204
202	17	280	144	302	173	86	1204
71	236	2	265	129	343	158	1204
1207	1204	1204	1204	1204	1204	1204	

121	335	150	70	228	43	257	1204
291	106	320	184	55	213	35	1204
20	276	147	305	169	89	198	1204
232	5	261	132	339	161	74	1204
59	224	39	246	117	324	195	1204
180	93	209	24	287	102	309	1204
301	165	78	243	9	272	136	1204
1204	1204	1204	1204	1204	1204	1204	

Y aspect

1	264	128	342	157	77	235	1204
84	242	8	271	135	300	164	1204
307	171	91	200	15	278	142	1204
285	100	314	178	98	207	22	1204
214	29	292	107	321	185	56	1204
192	63	221	36	250	114	328	1204
121	335	150	70	228	43	257	1204
1204	1204	1204	1204	1204	1204	1204	

165	78	243	9	272	136	301	1204
143	308	172	85	201	16	279	1204
23	286	101	315	179	92	208	1204
50	215	30	293	108	322	186	1204
329	193	57	222	37	251	115	1204
258	122	336	151	64	229	44	1204
236	2	265	129	343	158	71	1204
1204	1204	1204	1204	1204	1204	1204	

280	144	302	173	86	202	17	1204
209	24	287	102	309	180	93	1204
187	51	216	31	294	109	316	1204
116	323	194	58	223	38	252	1204
46	259	123	330	152	65	230	1205
72	237	3	266	130	337	159	1204
295	166	79	244	10	273	137	1204
1205	1204	1204	1204	1204	1204	1204	

94	210	25	281	103	310	181	1204
317	188	52	217	32	288	110	1204
246	117	324	195	59	224	39	1204
231	46	253	124	331	153	66	1204
160	73	238	4	260	131	338	1204
138	296	167	80	245	11	267	1204
18	274	145	303	174	87	203	1204
1204	1204	1204	1204	1204	1204	1204	

111	318	189	53	211	33	289	1204
40	247	118	325	196	60	218	1204
67	225	47	254	125	332	154	1204
339	161	74	232	5	261	132	1204
268	139	297	168	81	239	12	1204
197	19	275	146	304	175	88	1204
182	95	204	26	282	104	311	1204
1204	1204	1204	1204	1204	1204	1204	

219	41	248	119	326	190	61	1204
148	68	226	48	255	126	333	1204
133	340	155	75	233	6	262	1204
13	269	140	298	162	82	240	1204
89	198	20	276	147	305	169	1204
312	176	96	205	27	283	105	1204
290	112	319	183	54	212	34	1204
1204	1204	1204	1204	1204	1204	1204	

334	149	69	227	49	256	120	1204
263	127	341	156	76	234	7	1204
241	14	270	134	299	163	83	1204
170	90	199	21	277	141	306	1204
99	313	177	97	206	28	284	1204
35	291	106	320	184	55	213	1204
62	220	42	249	113	327	191	1204
1204	1204	1204	1204	1204	1204	1204	

Z aspect

Unbroken diagonals listed separately:

1	264	128	342	157	77	235	1204
279	153	27	342	216	139	62	1218
164	207	250	342	91	29	121	1204
334	318	302	342	326	310	301	2233
168	205	249	342	85	31	124	1204
4	266	129	342	156	75	232	1204
94	144	243	342	49	190	289	1351
218	82	184	342	101	259	18	1204

X aspect

1	264	128	342	157	77	235	1204
121	29	91	342	250	207	164	1204
301	310	326	342	302	318	334	2233
279	153	27	342	216	139	62	1218
285	171	8	342	228	114	56	1204
115	11	54	342	287	225	170	1204
168	205	249	342	85	31	124	1204
232	75	156	342	129	266	4	1204

Y aspect

1	264	128	342	157	77	235	1204
334	318	302	342	326	310	301	2233
62	139	216	342	27	153	279	1218
121	29	91	342	250	207	164	1204
94	144	243	342	157	77	235	1292
18	259	101	342	184	82	218	1204
285	171	8	342	228	114	56	1204
170	225	287	342	54	11	115	1204

Z aspect

APPENDIX IV

Diagonals of 8x8 cube Fig 48

112	450	19	445	501	91	394	40	2052
433	431	462	196	44	54	87	345	2052
480	114	179	285	69	491	298	136	2052
1	271	366	372	412	150	247	233	2052
384	210	259	173	229	331	154	312	2052
161	191	222	468	316	294	327	73	2052
208	354	419	13	341	251	58	408	2052
273	31	126	100	140	390	487	505	2052

32	127	99	141	389	486	506	280	2060
449	18	446	500	92	395	39	105	2044
432	463	195	45	53	86	346	440	2060
113	178	286	68	492	299	135	473	2044
272	367	371	413	149	246	234	8	2060
209	258	174	228	332	155	311	377	2044
192	223	467	317	293	326	74	168	2060
353	418	14	340	252	59	407	201	2044

417	15	339	253	60	406	202	360	2052
128	98	142	388	485	507	279	25	2052
17	447	499	93	396	38	106	456	2052
464	194	46	52	85	347	439	425	2052
177	287	67	493	300	134	474	120	2052
368	370	414	148	245	235	7	265	2052
257	175	227	333	156	310	378	216	2052
224	466	318	292	325	75	167	185	2052

465	319	291	324	76	166	186	217	2044
16	338	254	61	405	203	359	424	2060
97	143	387	484	508	278	26	121	2044
448	498	94	397	37	107	455	24	2060
193	47	51	84	348	438	426	457	2044
288	66	494	301	133	475	119	184	2060
369	415	147	244	236	6	266	361	2044
176	226	334	157	309	379	215	264	2060

225	335	158	308	380	214	263	169	2052
320	290	323	77	165	187	218	472	2052
337	255	62	404	204	358	423	9	2052
144	386	483	509	277	27	122	104	2052
497	95	398	36	108	454	23	441	2052
48	50	83	349	437	427	458	200	2052
65	495	302	132	476	118	183	281	2052
416	146	243	237	5	267	362	376	2052

145	242	238	4	268	363	375	409	2044
336	159	307	381	213	262	170	232	2060
289	322	78	164	188	219	471	313	2044
256	63	403	205	357	422	10	344	2060
385	482	510	276	28	123	103	137	2044
96	399	35	109	453	22	442	504	2060
49	82	350	436	428	459	199	41	2044
496	303	131	477	117	182	282	72	2060

304	130	478	116	181	283	71	489	2052
241	239	3	269	364	374	410	152	2052
160	306	382	212	261	171	231	329	2052
321	79	163	189	220	470	314	296	2052
64	402	206	356	421	11	343	249	2052
481	511	275	29	124	102	138	392	2052
400	34	110	452	21	443	503	89	2052
81	351	435	429	460	198	42	56	2052

352	434	430	461	197	43	55	88	2060
129	479	115	180	284	70	490	297	2044
240	2	270	365	373	411	151	248	2060
305	383	211	260	172	230	330	153	2044
80	162	190	221	469	315	295	328	2060
401	207	355	420	12	342	250	57	2044
512	274	30	125	101	139	391	488	2060
33	111	451	20	444	502	90	393	2044

X aspect

4

112	450	19	445	501	91	394	40	2052
32	127	99	141	389	486	506	280	2060
417	15	339	253	60	406	202	360	2052
465	319	291	324	76	166	186	217	2044
225	335	158	308	380	214	263	169	2052
145	242	238	4	268	363	375	409	2044
304	130	478	116	181	283	71	489	2052
352	434	430	461	197	43	55	88	2060

111	451	20	444	502	90	393	33	2044
31	126	100	140	390	487	505	273	2052
418	14	340	252	59	407	201	353	2044
466	318	292	325	75	167	185	224	2052
226	334	157	309	379	215	264	176	2060
146	243	237	5	267	362	376	416	2052
303	131	477	117	182	282	72	496	2060
351	435	429	460	198	42	56	81	2052

110	452	21	443	503	89	400	34	2052
30	125	101	139	391	488	512	274	2060
419	13	341	251	58	408	208	354	2052
467	317	293	326	74	168	192	223	2060
227	333	156	310	378	216	257	175	2052
147	244	236	6	266	361	369	415	2044
302	132	476	118	183	281	65	495	2052
350	436	428	459	199	41	49	82	2044

109	453	22	442	504	96	399	35	2060
29	124	102	138	392	481	511	275	2052
420	12	342	250	57	401	207	355	2044
468	316	294	327	73	161	191	222	2052
228	332	155	311	377	209	258	174	2044
148	245	235	7	265	368	370	414	2052
301	133	475	119	184	288	66	494	2060
349	437	427	458	200	48	50	83	2052

108	454	23	441	497	95	398	36	2052
28	123	103	137	385	482	510	276	2044
421	11	343	249	64	402	206	356	2052
469	315	295	328	80	162	190	221	2060
229	331	154	312	384	210	259	173	2052
149	246	234	8	272	367	371	413	2060
300	134	474	120	177	287	67	493	2052
348	438	426	457	193	47	51	84	2044

107	455	24	448	498	94	397	37	2060
27	122	104	144	386	483	509	277	2052
422	10	344	256	63	403	205	357	2060
470	314	296	321	79	163	189	220	2052
230	330	153	305	383	211	260	172	2044
150	247	233	1	271	366	372	412	2052
299	135	473	113	178	286	68	492	2044
347	439	425	464	194	46	52	85	2052

106	456	17	447	499	93	396	38	2052
26	121	97	143	387	484	508	278	2044
423	9	337	255	62	404	204	358	2052
471	313	289	322	78	164	188	219	2044
231	329	160	306	382	212	261	171	2052
151	248	240	2	270	365	373	411	2060
298	136	480	114	179	285	69	491	2052
346	440	432	463	195	45	53	86	2060

105	449	18	446	500	92	395	39	2044
25	128	98	142	388	485	507	279	2052
424	16	338	254	61	405	203	359	2060
472	320	290	323	77	165	187	218	2052
232	336	159	307	381	213	262	170	2060
152	241	239	3	269	364	374	410	2052
297	129	479	115	180	284	70	490	2044
345	433	431	462	196	44	54	87	2052

Y aspect

112	450	19	445	501	91	394	40	2052
111	451	20	444	502	90	393	33	2044
110	452	21	443	503	89	400	34	2052
109	453	22	442	504	96	399	35	2060
108	454	23	441	497	95	398	36	2052
107	455	24	448	498	94	397	37	2060
106	456	17	447	499	93	396	38	2052
106	456	17	447	499	93	396	38	2052

433	431	462	196	44	54	87	345	2052
434	430	461	197	43	55	88	352	2060
435	429	460	198	42	56	81	351	2052
436	428	459	199	41	49	82	350	2044
437	427	458	200	48	50	83	349	2052
438	426	457	193	47	51	84	348	2044
439	425	464	194	46	52	85	347	2052
440	432	463	195	45	53	86	346	2060

480	114	179	285	69	491	298	136	2052
479	115	180	284	70	490	297	129	2044
478	116	181	283	71	489	304	130	2052
477	117	182	282	72	496	303	131	2060
476	118	183	281	65	495	302	132	2052
475	119	184	288	66	494	301	133	2060
474	120	177	287	67	493	300	134	2052
473	113	178	286	68	492	299	135	2044

1	271	366	372	412	150	247	233	2052
2	270	365	373	411	151	248	240	2060
3	269	364	374	410	152	241	239	2052
4	268	363	375	409	145	242	238	2044
5	267	362	376	416	146	243	237	2052
6	266	361	369	415	147	244	236	2044
7	265	368	370	414	148	245	235	2052
8	272	367	371	413	149	246	234	2060

384	210	259	173	229	331	154	312	2052
383	211	260	172	230	330	153	305	2044
382	212	261	171	231	329	160	306	2052
381	213	262	170	232	336	159	307	2060
380	214	263	169	225	335	158	308	2052
379	215	264	176	226	334	157	309	2060
378	216	257	175	227	333	156	310	2052
377	209	258	174	228	332	155	311	2044

161	191	222	468	316	294	327	73	2052
162	190	221	469	315	295	328	80	2060
163	189	220	470	314	296	321	79	2052
164	188	219	471	313	289	322	78	2044
165	187	218	472	320	290	323	77	2052
166	186	217	465	319	291	324	76	2044
167	185	224	466	318	292	325	75	2052
168	192	223	467	317	293	326	74	2060

208	354	419	13	341	251	58	408	2052
207	355	420	12	342	250	57	401	2044
206	356	421	11	343	249	64	402	2052
205	357	422	10	344	256	63	403	2060
204	358	423	9	337	255	62	404	2052
203	359	424	16	338	254	61	405	2060
202	360	417	15	339	253	60	406	2052
201	353	418	14	340	252	59	407	2044

273	31	126	100	140	390	487	505	2052
274	30	125	101	139	391	488	512	2060
275	29	124	102	138	392	481	511	2052
276	28	123	103	137	385	482	510	2044
277	27	122	104	144	386	483	509	2052
278	26	121	97	143	387	484	508	2044
279	25	128	98	142	388	485	507	2052
280	32	127	99	141	389	486	506	2060

Z aspect

Unbroken diagonals listed separately:

112	450	19	445	501	91	394	40	2052
352	242	291	141	197	363	186	280	2052
33	399	94	500	444	22	455	105	2052
273	191	366	196	140	294	247	345	2052

X aspect

112	450	19	445	501	91	394	40	2052
105	455	22	444	500	94	399	33	2052
345	247	294	140	196	366	191	273	2052
352	242	291	141	197	363	186	280	2052

Y aspect

112	450	19	445	501	91	394	40	2052
273	191	366	196	140	294	247	345	2052
280	186	363	197	141	291	242	352	2052
105	455	22	444	500	94	399	33	2052

Z aspect

APPENDIX V

Diagonals of 9x9 cube Fig 51

130	347	584	488	43	276	477	705	235	3285
314	461	727	213	108	336	568	526	32	3285
331	609	522	66	298	481	698	188	92	3285
459	669	172	85	365	602	551	61	321	3285
631	544	77	296	425	664	168	126	354	3285
653	206	110	394	627	567	48	262	418	3285
533	25	258	414	687	190	148	383	647	3285
231	144	399	613	508	14	251	443	682	3285
3	280	436	716	224	155	376	591	504	3285

685	193	149	386	641	538	24	261	408	3285
509	17	245	448	681	234	138	397	616	3285
218	160	375	594	498	1	283	437	719	3285
42	279	471	703	238	131	350	578	493	3285
102	334	571	527	35	308	466	726	216	3285
301	482	701	182	97	330	612	516	64	3285
368	596	556	60	324	453	667	175	86	3285
430	663	171	120	352	634	545	80	290	3285
630	561	46	265	419	656	200	115	393	3285

318	451	670	176	89	362	601	555	63	3285
355	635	548	74	295	429	666	165	118	3285
422	650	205	114	396	624	559	49	266	3285
646	537	27	255	406	688	194	152	380	3285
684	228	136	400	617	512	11	250	447	3285
496	4	284	440	713	223	159	378	588	3285
239	134	344	583	492	45	273	469	706	3285
29	313	465	729	210	100	337	572	530	3285
96	333	606	514	67	302	485	695	187	3285

495	39	271	472	707	242	128	349	582	3285
208	103	338	575	524	34	312	468	723	3285
68	305	479	700	186	99	327	604	517	3285
83	367	600	558	57	316	454	671	179	3285
294	432	660	163	121	356	638	542	79	3285
390	622	562	50	269	416	655	204	117	3285
409	689	197	146	385	645	540	21	253	3285
620	506	16	249	450	678	226	139	401	3285
718	222	162	372	586	499	5	287	434	3285

384	648	534	19	256	410	692	191	151	3285
444	676	229	140	404	614	511	15	252	3285
589	500	8	281	439	717	225	156	370	3285
710	236	133	348	585	489	37	274	473	3285
529	33	315	462	721	211	104	341	569	3285
189	93	325	607	518	71	299	484	699	3285
55	319	455	674	173	88	366	603	552	3285
122	359	632	547	78	297	426	658	166	3285
263	421	654	207	111	388	625	563	53	3285

178	87	369	597	550	58	320	458	668	3285
81	291	424	661	167	125	353	637	546	3285
109	391	626	566	47	268	420	657	201	3285
257	413	686	196	150	387	642	532	22	3285
398	619	510	18	246	442	679	230	143	3285
438	720	219	154	373	590	503	2	286	3285
579	487	40	275	476	704	241	132	351	3285
724	212	107	335	574	528	36	309	460	3285
521	65	304	483	702	183	91	328	608	3285

470	709	240	135	345	577	490	41	278	3285
573	531	30	307	463	725	215	101	340	3285
696	181	94	329	611	515	70	303	486	3285
553	59	323	452	673	177	90	363	595	3285
170	119	358	636	549	75	289	427	662	3285
52	267	423	651	199	112	392	629	560	3285
153	381	640	535	23	260	407	691	195	3285
244	445	680	233	137	403	618	513	12	3285
374	593	497	7	285	441	714	217	157	3285

26	254	412	690	198	147	379	643	536	3285
142	402	621	507	10	247	446	683	227	3285
288	435	712	220	158	377	587	502	6	3285
343	580	491	44	272	475	708	243	129	3285
464	728	209	106	339	576	525	28	310	3285
605	520	69	306	480	694	184	95	332	3285
672	180	84	361	598	554	62	317	457	3285
543	73	292	428	665	164	124	357	639	3285
202	113	395	623	565	51	270	417	649	3285

599	557	56	322	456	675	174	82	364	3285
659	169	123	360	633	541	76	293	431	3285
564	54	264	415	652	203	116	389	628	3285
192	145	382	644	539	20	259	411	693	3285
13	248	449	677	232	141	405	615	505	3285
161	371	592	501	9	282	433	715	221	3285
277	474	711	237	127	346	581	494	38	3285
342	570	523	31	311	467	722	214	105	3285
478	697	185	98	326	610	519	72	300	3285

X aspect

130	347	584	488	43	276	477	705	235	3285
685	193	149	386	641	538	24	261	408	3285
318	451	670	176	89	362	601	555	63	3285
495	39	271	472	707	242	128	349	582	3285
384	648	534	19	256	410	692	191	151	3285
178	87	369	597	550	58	320	458	668	3285
470	709	240	135	345	577	490	41	278	3285
26	254	412	690	198	147	379	643	536	3285
599	557	56	322	456	675	174	82	364	3285

697	185	98	326	610	519	72	300	478	3285
280	436	716	224	155	376	591	504	3	3285
561	46	265	419	656	200	115	393	630	3285
333	606	514	67	302	485	695	187	96	3285
222	162	372	586	499	5	287	434	718	3285
421	654	207	111	388	625	563	53	263	3285
65	304	483	702	183	91	328	608	521	3285
593	497	7	285	441	714	217	157	374	3285
113	395	623	565	51	270	417	649	202	3285

292	428	665	164	124	357	639	543	73	3285
523	31	311	467	722	214	105	342	570	3285
399	613	508	14	251	443	682	231	144	3285
171	120	352	634	545	80	290	430	663	3285
465	729	210	100	337	572	530	29	313	3285
16	249	450	678	226	139	401	620	506	3285
632	547	78	297	426	658	166	122	359	3285
107	335	574	528	36	309	460	724	212	3285
680	233	137	403	618	513	12	244	445	3285

535	23	260	407	691	195	153	381	640	3285
361	598	554	62	317	457	672	180	84	3285
237	127	346	581	494	38	277	474	711	3285
414	687	190	148	383	647	533	25	258	3285
60	324	453	667	175	86	368	596	556	3285
583	492	45	273	469	706	239	134	344	3285
146	385	645	540	21	253	409	689	197	3285
674	173	88	366	603	552	55	319	455	3285
275	476	704	241	132	351	579	487	40	3285

373	590	503	2	286	438	720	219	154	3285
199	112	392	629	560	52	267	423	651	3285
480	694	184	95	332	605	520	69	306	3285
9	282	433	715	221	161	371	592	501	3285
627	567	48	262	418	653	206	110	394	3285
97	330	612	516	64	301	482	701	182	3285
713	223	159	378	588	496	4	284	440	3285
269	416	655	204	117	390	622	562	50	3285
518	71	299	484	699	189	93	325	607	3285

211	104	341	569	529	33	315	462	721	3285
442	679	230	143	398	619	510	18	246	3285
75	289	427	662	170	119	358	636	549	3285
576	525	28	310	464	728	209	106	339	3285
141	405	615	505	13	248	449	677	232	3285
664	168	126	354	631	544	77	296	425	3285
308	466	726	216	102	334	571	527	35	3285
512	11	250	447	684	228	136	400	617	3285
356	638	542	79	294	432	660	163	121	3285

454	671	179	83	367	600	558	57	316	3285
37	274	473	710	236	133	348	585	489	3285
642	532	22	257	413	686	196	150	387	3285
90	363	595	553	59	323	452	673	177	3285
708	243	129	343	580	491	44	272	475	3285
259	411	693	192	145	382	644	539	20	3285
551	61	321	459	669	172	85	365	602	3285
350	578	493	42	279	471	703	238	131	3285
194	152	380	646	537	27	255	406	688	3285

49	266	422	650	205	114	396	624	559	3285
604	517	68	305	479	700	186	99	327	3285
156	370	589	500	8	281	439	717	225	3285
657	201	109	391	626	566	47	268	420	3285
303	486	696	181	94	329	611	515	70	3285
502	6	288	435	712	220	158	377	587	3285
389	628	564	54	264	415	652	203	116	3285
188	92	331	609	522	66	298	481	698	3285
437	719	218	160	375	594	498	1	283	3285

616	509	17	245	448	681	234	138	397	3285
118	355	635	548	74	295	429	666	165	3285
723	208	103	338	575	524	34	312	468	3285
252	444	676	229	140	404	614	511	15	3285
546	81	291	424	661	167	125	353	637	3285
340	573	531	30	307	463	725	215	101	3285
227	142	402	621	507	10	247	446	683	3285
431	659	169	123	360	633	541	76	293	3285
32	314	461	727	213	108	336	568	526	3285

Y aspect

130	347	584	488	43	276	477	705	235	3285
697	185	98	326	610	519	72	300	478	3285
292	428	665	164	124	357	639	543	73	3285
535	23	260	407	691	195	153	381	640	3285
373	590	503	2	286	438	720	219	154	3285
211	104	341	569	529	33	315	462	721	3285
454	671	179	83	367	600	558	57	316	3285
49	266	422	650	205	114	396	624	559	3285
616	509	17	245	448	681	234	138	397	3285

314	461	727	213	108	336	568	526	32	3285
557	56	322	456	675	174	82	364	599	3285
395	623	565	51	270	417	649	202	113	3285
233	137	403	618	513	12	244	445	680	3285
476	704	241	132	351	579	487	40	275	3285
71	299	484	699	189	93	325	607	518	3285
638	542	79	294	432	660	163	121	356	3285
152	380	646	537	27	255	406	688	194	3285
719	218	160	375	594	498	1	283	437	3285

331	609	522	66	298	481	698	188	92	3285
169	123	360	633	541	76	293	431	659	3285
412	690	198	147	379	643	536	26	254	3285
7	285	441	714	217	157	74	593	497	2985
574	528	36	309	460	724	212	107	335	3285
88	366	603	552	55	319	455	674	173	3285
655	204	117	390	622	562	50	269	416	3285
250	447	684	228	136	400	617	512	11	3285
493	42	279	471	703	238	131	350	578	3285

459	669	172	85	365	602	551	61	321	3285
54	264	415	652	203	116	389	628	564	3285
621	507	10	247	446	683	227	142	402	3285
135	345	577	490	41	278	470	709	240	3285
702	183	91	328	608	521	65	304	483	3285
297	426	658	166	122	359	632	547	78	3285
540	21	253	409	689	197	146	385	645	3285
378	588	496	4	284	440	713	223	159	3285
216	102	334	571	527	35	308	466	726	3285

631	544	77	296	425	664	168	126	354	3285
145	382	644	539	20	259	411	693	192	3285
712	220	158	377	587	502	6	288	435	3285
307	463	725	215	101	340	573	531	30	3285
550	58	320	458	668	178	87	369	597	3285
388	625	563	53	263	421	654	207	111	3285
226	139	401	620	506	16	249	450	678	3285
469	706	239	134	344	583	492	45	273	3285
64	301	482	701	182	97	330	612	516	3285

653	206	110	394	627	567	48	262	418	3285
248	449	677	232	141	405	615	505	13	3285
491	44	272	475	708	243	129	343	580	3285
329	611	515	70	303	486	696	181	94	3285
167	125	353	637	546	81	291	424	661	3285
410	692	191	151	384	648	534	19	256	3285
5	287	434	718	222	162	372	586	499	3285
572	530	29	313	465	729	210	100	337	3285
86	368	596	556	60	324	453	667	175	3285

533	25	258	414	687	190	148	383	647	3285
371	592	501	9	282	433	715	221	161	3285
209	106	339	576	525	28	310	464	728	3285
452	673	177	90	363	595	553	59	323	3285
47	268	420	657	201	109	391	626	566	3285
614	511	15	252	444	676	229	140	404	3285
128	349	582	495	39	271	472	707	242	3285
695	187	96	333	606	514	67	302	485	3285
290	430	663	171	120	352	634	545	80	3285

231	144	399	613	508	14	251	443	682	3285
474	711	237	127	346	581	494	38	277	3285
69	306	480	694	184	95	332	605	520	3285
636	549	75	289	427	662	170	119	358	3285
150	387	642	532	22	257	413	686	196	3285
717	225	156	370	589	500	8	281	439	3285
312	468	723	208	103	338	575	524	34	3285
555	63	318	451	670	176	89	362	601	3285
393	630	561	46	265	419	656	200	115	3285

3	280	436	716	224	155	376	591	504	3285
570	523	31	311	467	722	214	105	342	3285
84	361	598	554	62	317	457	672	180	3285
651	199	112	392	629	560	52	267	423	3285
246	442	679	230	143	398	619	510	18	3285
489	37	274	473	710	236	133	348	585	3285
327	604	517	68	305	479	700	186	99	3285
165	118	355	633	548	76	295	429	666	3285
408	685	193	149	386	641	538	24	261	3285

Z aspect

Unbroken diagonals listed separately:

130	347	584	488	43	276	477	705	235	3285
599	709	534	176	43	147	320	349	408	3285
478	381	315	114	43	164	503	671	616	3285
3	25	77	66	43	14	48	61	32	369
384	39	670	386	43	675	379	41	668	3285
263	689	136	594	43	467	184	363	546	3285
631	669	522	213	43	155	251	383	418	3285
13	59	8	74	43	51	36	21	64	369

130	347	584	488	43	276	477	705	235	3285
616	671	503	164	43	114	315	381	478	3285
32	61	48	14	43	66	77	25	3	369
599	709	534	176	43	147	320	349	408	3285
373	23	665	326	43	681	396	57	721	3285
518	385	210	419	43	633	158	673	246	3285
384	39	670	386	43	675	379	41	668	3285
546	363	184	467	43	594	136	689	263	3285

130	347	584	488	43	276	477	705	235	3285
3	25	77	66	43	14	48	61	32	369
408	349	320	147	43	176	534	709	599	3285
616	671	503	164	43	114	315	381	478	3285
631	669	522	213	43	155	251	383	418	3285
64	21	36	51	43	74	8	59	13	369
373	23	665	326	43	681	396	57	721	3285
246	673	158	633	43	419	210	385	518	3285

X aspect   **Y aspect**   **Z aspect**

APPENDIX VI

Diagonals of 9x9 cube Fig No. 55

454	669	501	569	284	352	72	142	242	3285
535	588	258	326	41	109	234	466	728	3285
292	345	15	83	203	433	720	547	647	3285
49	102	177	407	689	514	639	304	404	3285
211	426	663	488	608	271	396	61	101	3285
697	507	582	245	365	28	153	223	485	3285
616	264	339	2	122	190	477	709	566	3285
373	21	96	164	446	676	558	628	323	3285
130	183	420	650	527	595	315	385	80	3285
3357	3105	3051	2934	3285	3168	3564	3465	3636	

5	118	198	475	710	562	615	267	335	3285
167	442	684	556	629	319	372	24	92	3285
653	523	603	313	386	76	129	186	416	3285
572	280	360	70	143	238	453	672	497	3285
329	37	117	232	467	724	534	591	254	3285
86	199	441	718	548	643	291	348	11	3285
410	685	522	637	305	400	48	105	173	3285
491	604	279	394	62	157	210	429	659	3285
248	361	36	151	224	481	696	510	578	3285
2961	3249	3240	3546	3474	3600	3348	3132	3015	

638	301	399	51	101	176	406	693	520	3285
395	58	156	213	425	662	487	612	277	3285
152	220	480	699	506	581	244	369	34	3285
476	706	561	618	263	338	1	126	196	3285
557	625	318	375	20	95	163	450	682	3285
314	382	75	132	182	419	649	531	601	3285
71	139	237	456	668	500	568	288	358	3285
233	463	723	537	587	257	325	45	115	3285
719	544	642	294	344	14	82	207	439	3285
3555	3438	3591	3375	3096	3042	2925	3321	3222	

452	671	496	576	286	359	67	138	240	3285
533	590	253	333	43	116	229	462	726	3285
290	347	10	90	205	440	715	543	645	3285
47	104	172	414	691	521	634	300	402	3285
209	428	658	495	610	278	391	57	159	3285
695	509	577	252	367	35	148	219	483	3285
614	266	334	9	124	197	472	705	564	3285
371	23	91	171	448	683	553	624	321	3285
128	185	415	657	529	602	310	381	78	3285
3339	3123	3006	2997	3303	3231	3519	3429	3618	

7	125	193	471	708	560	617	262	342	3285
169	449	679	552	627	317	374	19	99	3285
655	530	598	309	384	74	131	181	423	3285
574	287	355	66	141	236	455	667	504	3285
331	44	112	228	465	722	536	586	261	3285
88	206	436	714	546	641	293	343	18	3285
412	692	517	633	303	398	50	100	180	3285
493	611	274	390	60	155	212	424	666	3285
250	368	31	147	222	479	698	505	585	3285
2979	3312	3195	3510	3456	3582	3366	3087	3078	

636	299	401	46	108	178	413	688	516	3285
393	56	158	208	432	664	494	607	273	3285
150	218	482	694	513	583	251	364	30	3285
474	704	563	613	270	340	8	121	192	3285
555	623	320	370	27	97	170	445	678	3285
312	380	77	127	189	421	656	526	597	3285
69	137	239	451	675	502	575	283	354	3285
231	461	725	532	594	259	332	40	111	3285
717	542	644	289	351	16	89	202	435	3285
3537	3420	3609	3330	3159	3060	2988	3276	3186	

459	673	503	571	282	357	65	140	235	3285
540	592	260	328	39	114	227	464	721	3285
297	349	17	85	201	438	713	545	640	3285
54	106	179	409	687	519	632	302	397	3285
216	430	665	490	606	276	389	59	154	3285
702	511	584	247	363	33	146	221	478	3285
621	268	341	4	120	195	470	707	559	3285
378	25	98	166	444	681	551	626	316	3285
135	187	422	652	525	600	308	383	73	3285
3402	3141	3069	2952	3267	3213	3501	3447	3573	

3	123	191	473	703	567	619	269	337	3285
165	447	677	554	622	324	376	26	94	3285
651	528	596	311	379	81	133	188	418	3285
570	285	353	68	136	243	457	674	499	3285
327	42	110	230	460	729	538	593	256	3285
84	204	434	716	541	648	295	350	13	3285
408	690	515	635	298	405	52	107	175	3285
489	609	272	392	55	162	214	431	661	3285
246	366	29	149	217	486	700	512	580	3285
2943	3294	3177	3528	3411	3645	3384	3150	3033	

631	306	403	53	103	174	411	686	518	3285
388	63	160	215	427	660	492	605	275	3285
145	225	484	701	508	579	249	362	32	3285
469	711	565	620	265	336	6	119	194	3285
550	630	322	377	22	93	168	443	680	3285
307	387	79	134	184	417	654	524	599	3285
64	144	241	458	670	498	573	281	356	3285
226	468	727	539	589	255	330	38	113	3285
712	549	646	296	346	12	87	200	437	3285
3492	3483	3627	3393	3114	3024	2970	3258	3204	

X aspect

454	669	501	569	284	352	72	142	242	3285
549	646	296	346	12	87	200	437	712	3285
272	392	55	162	214	431	661	489	609	3285
4	120	195	470	707	559	621	268	341	3285
189	421	656	526	597	312	380	77	127	3285
722	536	586	261	331	44	112	228	465	3285
634	300	402	47	104	172	414	691	521	3285
369	34	152	220	480	699	506	581	244	3285
92	167	442	684	556	629	319	372	24	3285
3285	3285	3285	3285	3285	3285	3285	3285	3285	

535	588	258	326	41	109	234	466	728	3285
306	403	53	103	174	411	686	518	631	3285
29	149	217	486	700	512	580	246	366	3285
166	444	681	551	626	316	378	25	98	3285
675	502	575	283	354	69	137	239	451	3285
641	293	343	18	88	206	436	714	546	3285
391	57	159	209	428	658	495	610	278	3285
126	196	476	706	561	618	263	338	1	3285
416	653	523	603	313	386	76	129	186	3285
3285	3285	3285	3285	3285	3285	3285	3285	3285	

292	345	15	83	203	433	720	547	647	3285
63	160	215	427	660	492	605	275	388	3285
191	473	703	567	619	269	337	3	123	3285
652	525	600	308	383	73	135	187	422	3285
594	259	332	40	111	231	461	725	532	3285
398	50	100	180	412	692	517	633	303	3285
148	219	483	695	509	577	252	367	35	3285
450	682	557	625	318	375	20	95	163	3285
497	572	280	360	70	143	238	453	672	3285
3285	3285	3285	3285	3285	3285	3285	3285	3285	

49	102	177	407	689	514	639	304	404	3285
225	484	701	508	579	249	362	32	145	3285
677	554	622	324	376	26	94	165	447	3285
571	282	357	65	140	235	459	673	503	3285
351	16	89	202	435	717	542	644	289	3285
155	212	424	666	493	611	274	390	60	3285
472	705	564	614	266	334	9	124	197	3285
531	601	314	382	75	132	182	419	649	3285
254	329	37	117	232	467	724	534	591	3285
3285	3285	3285	3285	3285	3285	3285	3285	3285	

211	426	663	488	608	271	396	61	161	3285
711	565	620	265	336	6	119	194	469	3285
596	311	379	81	133	188	418	651	528	3285
328	39	114	227	464	721	540	592	260	3285
108	178	413	688	516	636	299	401	46	3285
479	698	505	585	250	368	31	147	222	3285
553	624	321	371	23	91	171	448	683	3285
288	358	71	139	237	456	668	500	568	3285
11	86	199	441	718	548	643	291	348	3285
3285	3285	3285	3285	3285	3285	3285	3285	3285	

697	507	582	245	365	28	153	223	485	3285
630	322	377	22	93	168	443	680	550	3285
353	68	136	243	457	674	499	570	285	3285
85	201	438	713	545	640	297	349	17	3285
432	664	494	607	273	393	56	158	208	3285
560	617	262	342	7	125	193	471	708	3285
310	381	78	128	185	415	657	529	602	3285
45	115	233	463	723	537	587	257	325	3285
173	410	685	522	637	305	400	48	105	3285
3285	3285	3285	3285	3285	3285	3285	3285	3285	

616	264	339	2	122	190	477	709	566	3285
387	79	134	184	417	654	524	599	307	3285
110	230	460	729	538	593	256	327	42	3285
409	687	519	632	302	397	54	106	179	3285
513	583	251	364	30	150	218	482	694	3285
317	374	19	99	169	449	679	552	627	3285
67	138	240	452	671	496	576	286	359	3285
207	439	719	544	642	294	344	14	82	3285
659	491	604	279	394	62	157	210	429	3285
3285	3285	3285	3285	3285	3285	3285	3285	3285	

373	21	96	164	446	676	558	628	323	3285
144	241	458	670	498	573	281	356	64	3285
434	716	541	648	295	350	13	84	204	3285
490	606	276	389	59	154	216	430	665	3285
270	340	8	121	192	474	704	563	613	3285
74	131	181	423	655	530	598	309	384	3285
229	462	726	533	590	253	333	43	116	3285
693	520	638	301	399	51	101	176	406	3285
578	248	361	36	151	224	481	696	510	3285
3285	3285	3285	3285	3285	3285	3285	3285	3285	

130	183	420	650	527	595	315	385	80	3285
468	727	539	589	255	330	38	113	226	3285
515	635	298	405	52	107	175	408	690	3285
247	363	33	146	221	478	702	511	584	3285
27	97	170	445	678	555	623	320	370	3285
236	455	667	504	574	287	355	66	141	3285
715	543	645	290	347	10	90	205	440	3285
612	277	395	58	156	213	425	662	487	3285
335	5	118	198	475	710	562	615	267	3285
3285	3285	3285	3285	3285	3285	3285	3285	3285	

Y aspect

454	669	501	569	284	352	72	142	242	3285
5	118	198	475	710	562	615	267	335	3285
638	301	399	51	101	176	406	693	520	3285
452	671	496	576	286	359	67	138	240	3285
7	125	193	471	708	560	617	262	342	3285
636	299	401	46	108	178	413	688	516	3285
459	673	503	571	282	357	65	140	235	3285
3	123	191	473	703	567	619	269	337	3285
631	306	403	53	103	174	411	686	518	3285
3285	3285	3285	3285	3285	3285	3285	3285	3285	

549	646	296	346	12	87	200	437	712	3285
183	420	650	527	595	315	385	80	130	3285
361	36	151	224	481	696	510	578	248	3285
544	642	294	344	14	82	207	439	719	3285
185	415	657	529	602	310	381	78	128	3285
368	31	147	222	479	698	505	585	250	3285
542	644	289	351	16	89	202	435	717	3285
187	422	652	525	600	308	383	73	135	3285
366	29	149	217	486	700	512	580	246	3285
3285	3285	3285	3285	3285	3285	3285	3285	3285	

272	392	55	162	214	431	661	489	609	3285
727	539	589	255	330	38	113	226	468	3285
96	164	446	676	558	628	323	373	21	3285
279	394	62	157	210	429	659	491	604	3285
723	537	587	257	325	45	115	233	463	3285
91	171	448	683	553	624	321	371	23	3285
274	390	60	155	212	424	666	493	611	3285
725	532	594	259	332	40	111	231	461	3285
98	166	444	681	551	626	316	378	25	3285
3285	3285	3285	3285	3285	3285	3285	3285	3285	

4	120	195	470	707	559	621	268	341	3285
635	298	405	52	107	175	408	690	515	3285
458	670	498	573	281	356	64	144	241	3285
2	122	190	477	709	566	616	264	339	3285
637	305	400	48	105	173	410	685	522	3285
456	668	500	568	288	358	71	139	237	3285
9	124	197	472	705	564	614	266	334	3285
633	303	398	50	100	180	412	692	517	3285
451	675	502	575	283	354	69	137	239	3285
3285	3285	3285	3285	3285	3285	3285	3285	3285	

189	421	656	526	597	312	380	77	127	3285
363	33	146	221	478	702	511	584	247	3285
541	648	295	350	13	84	204	434	716	3285
184	417	654	524	599	307	387	79	134	3285
365	28	153	223	485	697	507	582	245	3285
548	643	291	348	11	86	199	441	718	3285
182	419	649	531	601	314	382	75	132	3285
367	35	148	219	483	695	509	577	252	3285
546	641	293	343	18	88	206	436	714	3285
3285	3285	3285	3285	3285	3285	3285	3285	3285	

722	536	586	261	331	44	112	228	465	3285
97	170	445	678	555	623	320	370	27	3285
276	389	59	154	216	430	665	490	606	3285
729	538	593	256	327	42	110	230	460	3285
93	168	443	680	550	630	322	377	22	3285
271	396	61	161	211	426	663	488	608	3285
724	534	591	254	329	37	117	232	467	3285
95	163	450	682	557	625	318	375	20	3285
278	391	57	159	209	428	658	495	610	3285
3285	3285	3285	3285	3285	3285	3285	3285	3285	

634	300	402	47	104	172	414	691	521	3285
455	667	504	574	287	355	66	141	236	3285
8	121	192	474	704	563	613	270	340	3285
632	302	397	54	106	179	409	687	519	3285
457	674	499	570	285	353	68	136	243	3285
6	119	194	469	711	565	620	265	336	3285
639	304	404	49	102	177	407	689	514	3285
453	672	497	572	280	360	70	143	238	3285
1	126	196	476	706	561	618	263	338	3285
3285	3285	3285	3285	3285	3285	3285	3285	3285	

369	34	152	220	480	699	506	581	244	3285
543	645	290	347	10	90	205	440	715	3285
181	423	655	530	598	309	384	74	131	3285
364	30	150	218	482	694	513	583	251	3285
545	640	297	349	17	85	201	438	713	3285
188	418	651	528	596	311	379	81	133	3285
362	32	145	225	484	701	508	579	249	3285
547	647	292	345	15	83	203	433	720	3285
186	416	653	523	603	313	386	76	129	3285
3285	3285	3285	3285	3285	3285	3285	3285	3285	

92	167	442	684	556	629	319	372	24	3285
277	395	58	156	213	425	662	487	612	3285
726	533	590	253	333	43	116	229	462	3285
99	169	449	679	552	627	317	374	19	3285
273	393	56	158	208	432	664	494	607	3285
721	540	592	260	328	39	114	227	464	3285
94	165	447	677	554	622	324	376	26	3285
275	388	63	160	215	427	660	492	605	3285
728	535	588	258	326	41	109	234	466	3285
3285	3285	3285	3285	3285	3285	3285	3285	3285	

Z aspect

Unbroken diagonals listed separately:

454	669	501	569	284	352	72	142	242	3285
92	300	656	162	284	699	112	268	712	3285
728	304	153	676	284	83	663	264	130	3285
631	673	193	51	284	567	413	138	335	3285
189	120	55	346	284	629	506	691	465	3285
546	124	587	224	284	427	379	687	27	3285
7	671	399	475	284	174	619	140	516	3285
273	302	295	255	284	313	318	266	250	2556

454	669	501	569	284	352	72	142	242	3285
130	264	663	83	284	676	153	304	728	3285
335	138	413	567	284	51	193	673	631	3285
92	300	656	162	284	699	112	268	712	3285
211	102	15	326	284	595	558	709	485	3285
11	705	332	486	284	213	598	106	550	3285
189	120	55	346	284	629	506	691	465	3285
27	687	379	427	284	224	587	124	546	3285

454	669	501	569	284	352	72	142	242	3285
631	673	193	51	284	567	413	138	335	3285
712	268	112	699	284	162	656	300	92	3285
130	264	663	83	284	676	153	304	728	3285
7	671	399	475	284	174	619	140	516	3285
250	266	318	313	284	255	295	302	273	2556
211	102	15	326	284	595	558	709	485	3285
550	106	598	213	284	486	332	705	11	3285

APPENDIX VII

THE SIXTEEN MORPHINGS OF FIG 28

	Rotatable Cube	X ASPECT		Diagonals

C AC AC C
AX 0 0 0 0

1	57	56	16	130
48	24	25	33	130
28	36	45	21	130
53	13	4	60	130
130	130	130	130	

62	6	11	51	130
19	43	38	30	130
34	26	23	47	130
15	55	58	2	130
130	130	130	130	

63	7	10	50	130
18	42	39	31	130
35	27	22	46	130
14	54	59	3	130
130	130	130	130	

5	61	52	12	130
44	20	29	37	130
32	40	41	17	130
49	9	8	64	130
130	130	130	130	

1	43	22	64	130
16	38	27	49	130
60	23	42	5	130
53	26	39	12	130

BX 0 90 90 0

1	57	56	16	130
48	24	25	33	130
28	36	45	21	130
53	13	4	60	130
130	130	130	130	

51	30	47	2	130
11	38	23	58	130
6	43	26	55	130
62	19	34	15	130
130	130	130	130	

50	31	46	3	130
10	39	22	59	130
7	42	27	54	130
63	18	35	14	130
130	130	130	130	

5	61	52	12	130
44	20	29	37	130
32	40	41	17	130
49	9	8	64	130
130	130	130	130	

1	38	27	64	130
16	23	42	49	130
60	26	39	5	130
53	43	22	12	130

CX 0 180 180 0

1	57	56	16	130
48	24	25	33	130
28	36	45	21	130
53	13	4	60	130
130	130	130	130	

2	58	55	15	130
47	23	26	34	130
30	38	43	19	130
51	11	6	62	130
130	130	130	130	

3	59	54	14	130
46	22	27	35	130
31	39	42	18	130
50	10	7	63	130
130	130	130	130	

5	61	52	12	130
44	20	29	37	130
32	40	41	17	130
49	9	8	64	130
130	130	130	130	

1	23	42	64	130
16	26	39	49	130
60	43	22	5	130
53	38	27	12	130

DX 0 270 270 0

1	57	56	16	130
48	24	25	33	130
28	36	45	21	130
53	13	4	60	130
130	130	130	130	

15	34	19	62	130
55	26	43	6	130
58	23	38	11	130
2	47	30	51	130
130	130	130	130	

14	35	18	63	130
54	27	42	7	130
59	22	39	10	130
3	46	31	50	130
130	130	130	130	

5	61	52	12	130
44	20	29	37	130
32	40	41	17	130
49	9	8	64	130
130	130	130	130	

1	26	39	64	130
16	43	22	49	130
60	38	27	5	130
53	23	42	12	130

EX 90 0 0 90

53	28	48	1	130
13	36	24	57	130
4	45	25	56	130
60	21	33	16	130
130	130	130	130	

62	6	11	51	130
19	43	38	30	130
34	26	23	47	130
15	55	58	2	130
130	130	130	130	

63	7	10	50	130
18	42	39	31	130
35	27	22	46	130
14	54	59	3	130
130	130	130	130	

49	32	44	5	130
9	40	20	61	130
8	41	29	52	130
64	17	37	12	130
130	130	130	130	

53	43	22	12	130
1	38	27	64	130
16	23	42	49	130
60	26	39	5	130

FX 90 90 90 90

53	28	48	1	130
13	36	24	57	130
4	45	25	56	130
60	21	33	16	130
130	130	130	130	

51	30	47	2	130
11	38	23	58	130
6	43	26	55	130
62	19	34	15	130
130	130	130	130	

50	31	46	3	130
10	39	22	59	130
7	42	27	54	130
63	18	35	14	130
130	130	130	130	

49	32	44	5	130
9	40	20	61	130
8	41	29	52	130
64	17	37	12	130
130	130	130	130	

53	38	27	12	130
1	23	42	64	130
16	26	39	49	130
60	38	27	5	130

GX 90 180 180 90

53	28	48	1	130
13	36	24	57	130
4	45	25	56	130
60	21	33	16	130
130	130	130	130	

2	58	55	15	130
47	23	26	34	130
30	38	43	19	130
51	11	6	62	130
130	130	130	130	

3	59	54	14	130
46	22	27	35	130
31	39	42	18	130
50	10	7	63	130
130	130	130	130	

49	32	44	5	130
9	40	20	61	130
8	41	29	52	130
64	17	37	12	130
130	130	130	130	

53	23	42	12	130
1	26	39	64	130
16	43	22	49	130
60	38	27	5	130

HX 90 270 270 90

53	28	48	1	130
13	36	24	57	130
4	45	25	56	130
60	21	33	16	130
130	130	130	130	

15	34	19	62	130
55	26	43	6	130
58	23	38	11	130
2	47	30	51	130
130	130	130	130	

14	35	18	63	130
54	27	42	7	130
59	22	39	10	130
3	46	31	50	130
130	130	130	130	

49	32	44	5	130
9	40	20	61	130
8	41	29	52	130
64	17	37	12	130
130	130	130	130	520

53	26	39	12	130
1	43	22	64	130
16	38	27	49	130
60	23	42	5	130

IX 180 0 0 180

60	4	13	53	130
21	45	36	28	130
33	25	24	48	130
16	56	57	1	130
130	130	130	130	

62	6	11	51	130
19	43	38	30	130
34	26	23	47	130
15	55	58	2	130
130	130	130	130	

63	7	10	50	130
18	42	39	31	130
35	27	22	46	130
14	54	59	3	130
130	130	130	130	

64	8	9	49	130
17	41	40	32	130
37	29	20	44	130
12	52	61	5	130
130	130	130	130	

60	43	22	5	130
53	38	27	12	130
1	23	42	64	130
16	26	39	49	130

JX 180 90 90 180

60	4	13	53	130
21	45	36	28	130
33	25	24	48	130
16	56	57	1	130
130	130	130	130	

51	30	47	2	130
11	38	23	58	130
6	43	26	55	130
62	19	34	15	130
130	130	130	130	

50	31	46	3	130
10	39	22	59	130
7	42	27	54	130
63	18	35	14	130
130	130	130	130	

64	8	9	49	130
17	41	40	32	130
37	29	20	44	130
12	52	61	5	130
130	130	130	130	

60	38	27	5	130
53	23	42	12	130
1	26	39	64	130
16	43	22	49	130

KX 180 180 180 180

60	4	13	53	130
21	45	36	28	130
33	25	24	48	130
16	56	57	1	130
130	130	130	130	

2	58	55	15	130
47	23	26	34	130
30	38	43	19	130
51	11	6	62	130
130	130	130	130	

3	59	54	14	130
46	22	27	35	130
31	39	42	18	130
50	10	7	63	130
130	130	130	130	

64	8	9	49	130
17	41	40	32	130
37	29	20	44	130
12	52	61	5	130
130	130	130	130	

60	23	42	5	130
53	26	39	12	130
1	43	22	64	130
16	38	27	49	130

LX 180 270 270 180

60	4	13	53	130
21	45	36	28	130
33	25	24	48	130
16	56	57	1	130
130	130	130	130	

15	34	19	62	130
55	26	43	6	130
58	23	38	11	130
2	47	30	51	130
130	130	130	130	

14	35	18	63	130
54	27	42	7	130
59	22	39	10	130
3	46	31	50	130
130	130	130	130	

64	8	9	49	130
17	41	40	32	130
37	29	20	44	130
12	52	61	5	130
130	130	130	130	

60	26	39	5	130
53	43	22	12	130
1	38	27	64	130
16	23	42	49	130

MX 270 0 0 270

16	33	21	60	130
56	25	45	4	130
57	24	36	13	130
1	48	28	53	130
130	130	130	130	

62	6	11	51	130
19	43	38	30	130
34	26	23	47	130
15	55	58	2	130
130	130	130	130	

63	7	10	50	130
18	42	39	31	130
35	27	22	46	130
14	54	59	3	130
130	130	130	130	

12	37	17	64	130
52	29	41	8	130
61	20	40	9	130
5	44	32	49	130
130	130	130	130	

16	43	22	49	130
60	38	27	5	130
53	23	42	12	130
1	26	39	64	130

NX 270 90 90 270

16	33	21	60	130
56	25	45	4	130
57	24	36	13	130
1	48	28	53	130
130	130	130	130	

51	30	47	2	130
11	38	23	58	130
6	43	26	55	130
62	19	34	15	130
130	130	130	130	

50	31	46	3	130
10	39	22	59	130
7	42	27	54	130
63	18	35	14	130
130	130	130	130	

12	37	17	64	130
52	29	41	8	130
61	20	40	9	130
5	44	32	49	130
130	130	130	130	

16	38	27	49	130
60	23	42	5	130
53	26	39	12	130
1	43	22	64	130

OX 170 180 180 270

16	33	21	60	130
56	25	45	4	130
57	24	36	13	130
1	48	28	53	130
130	130	130	130	

2	58	55	15	130
47	23	26	34	130
30	38	43	19	130
51	11	6	62	130
130	130	130	130	

3	59	54	14	130
46	22	27	35	130
31	39	42	18	130
50	10	7	63	130
130	130	130	130	

12	37	17	64	130
52	29	41	8	130
61	20	40	9	130
5	44	32	49	130
130	130	130	130	

16	23	42	49	130
60	26	39	5	130
53	43	22	12	130
1	38	27	64	130

PX 270 270 270 270

16	33	21	60	130
56	25	45	4	130
57	24	36	13	130
1	48	28	53	130
130	130	130	130	

15	34	19	62	130
55	26	43	6	130
58	23	38	11	130
2	47	30	51	130
130	130	130	130	

14	35	18	63	130
54	27	42	7	130
59	22	39	10	130
3	46	31	50	130
130	130	130	130	

12	37	17	64	130
52	29	41	8	130
61	20	40	9	130
5	44	32	49	130
130	130	130	130	

16	26	39	49	130
60	43	22	5	130
53	38	27	12	130
1	23	42	64	130

	Rotatable Cube	Y ASPECT	Diagonals

C AC AC C

AY 0 0 0 0

Rotatable Cube:

1	48	28	53	130
62	19	34	15	130
63	18	35	14	130
5	44	32	49	130
131	129	129	131	

57	24	36	13	130
6	43	26	55	130
7	42	27	54	130
61	20	40	9	130
131	129	129	131	

Y ASPECT:

56	25	45	4	130
11	38	23	58	130
10	39	22	59	130
52	29	41	8	130
129	131	131	129	

16	33	21	60	130
51	30	47	2	130
50	31	46	3	130
12	37	17	64	130
129	131	131	129	

Diagonals:

1	43	22	64	130	
53	26	39	12	130	P
49	27	38	16	130	
5	42	23	60	130	

BY 0 90 90 0

Rotatable Cube:

1	48	28	53	130
62	19	34	15	130
63	18	35	14	130
5	44	32	49	130
131	129	129	131	

13	55	54	9	131
36	26	27	40	129
24	43	42	20	129
57	6	7	61	131
130	130	130	130	

Y ASPECT:

4	58	59	8	129
45	23	22	41	131
25	38	39	29	131
56	11	10	52	129
130	130	130	130	

16	33	21	60	130
51	30	47	2	130
50	31	46	3	130
12	37	17	64	130
129	131	131	129	

Diagonals:

1	26	39	64	130
53	27	38	12	130
49	42	23	16	130
5	43	22	60	130

CY 0 180 180 0

Rotatable Cube:

1	48	28	53	130
62	19	34	15	130
63	18	35	14	130
5	44	32	49	130
131	129	129	131	

9	40	20	61	130
54	27	42	7	130
55	26	43	6	130
13	36	24	57	130
131	129	129	131	

Y ASPECT:

8	41	29	52	130
59	22	39	10	130
58	23	38	11	130
4	45	25	56	130
129	131	131	129	

16	33	21	60	130
51	30	47	2	130
50	31	46	3	130
12	37	17	64	130
129	131	131	129	

Diagonals:

1	27	38	64	130
53	42	23	12	130
49	43	22	16	130
5	26	39	60	130

DY 0 270 270 0

Rotatable Cube:

1	48	28	53	130
62	19	34	15	130
63	18	35	14	130
5	44	32	49	130
131	129	129	131	

61	7	6	57	131
20	42	43	24	129
40	27	26	36	129
9	54	55	13	131
130	130	130	130	

Y ASPECT:

52	10	11	56	129
29	39	38	25	131
41	22	23	45	131
8	59	58	4	129
130	130	130	130	

16	33	21	60	130
51	30	47	2	130
50	31	46	3	130
12	37	17	64	130
129	131	131	129	

Diagonals:

1	42	23	64	130
53	43	22	12	130
49	26	39	16	130
5	27	38	60	130

EY 90 0 0 90

Rotatable Cube:

5	63	62	1	131
44	18	19	48	129
32	35	34	28	129
49	14	15	53	131
130	130	130	130	

57	24	36	13	130
6	43	26	55	130
7	42	27	54	130
61	20	40	9	130
131	129	129	131	

Y ASPECT:

56	25	45	4	130
11	38	23	58	130
10	39	22	59	130
52	29	41	8	130
129	131	131	129	

12	50	51	16	129
37	31	30	33	131
17	46	47	21	131
64	3	2	60	129
130	130	130	130	

Diagonals:

5	43	22	60	130
1	26	39	64	130
53	27	38	12	130
49	42	23	16	130

FY 90 90 90 90

Rotatable Cube:

5	63	62	1	131
44	18	19	48	129
32	35	34	28	129
49	14	15	53	131
130	130	130	130	

13	55	54	9	131
36	26	27	40	129
24	43	42	20	129
57	6	7	61	131
130	130	130	130	

Y ASPECT:

4	58	59	8	129
45	23	22	41	131
25	38	39	29	131
56	11	10	52	129
130	130	130	130	

12	50	51	16	129
37	31	30	33	131
17	46	47	21	131
64	3	2	60	129
130	130	130	130	

Diagonals:

5	26	39	60	130
1	27	38	64	130
53	42	23	12	130
49	43	22	16	130

GY 90 180 180 90

Rotatable Cube:

5	63	62	1	131
44	18	19	48	129
32	35	34	28	129
49	14	15	53	131
130	130	130	130	

9	40	20	61	130
54	27	42	7	130
55	26	43	6	130
13	36	24	57	130
131	129	129	131	

Y ASPECT:

8	41	29	52	130
59	22	39	10	130
58	23	38	11	130
4	45	25	56	130
129	131	131	129	

12	50	51	16	129
37	31	30	33	131
17	46	47	21	131
64	3	2	60	129
130	130	130	130	

Diagonals:

5	27	38	60	130
1	42	23	64	130
53	43	22	12	130
49	26	39	16	130

HY 90 270 270 90

Rotatable Cube:

5	63	62	1	131
44	18	19	48	129
32	35	34	28	129
49	14	15	53	131
130	130	130	130	

61	7	6	57	131
20	42	43	24	129
40	27	26	36	129
9	54	55	13	131
130	130	130	130	

Y ASPECT:

52	10	11	56	129
29	39	38	25	131
41	22	23	45	131
8	59	58	4	129
130	130	130	130	

12	50	51	16	129
37	31	30	33	131
17	46	47	21	131
64	3	2	60	129
130	130	130	130	

Diagonals:

5	42	23	60	130	
1	43	22	64	130	P
53	26	39	12	130	
49	27	38	16	130	

144

IY 180 0 0 180

49	32	44	5	130
14	35	18	63	130
15	34	19	62	130
53	28	48	1	130
131	129	129	131	

57	24	36	13	130
6	43	26	55	130
7	42	27	54	130
61	20	40	9	130
131	129	129	131	

56	25	45	4	130
11	38	23	58	130
10	39	22	59	130
52	29	41	8	130
129	131	131	129	

64	17	37	12	130
3	46	31	50	130
2	47	30	51	130
60	21	33	16	130
129	131	131	129	

49	43	22	16	130
5	26	39	60	130
1	27	38	64	130
53	42	23	12	130

JY 180 90 90 180

49	32	44	5	130
14	35	18	63	130
15	34	19	62	130
53	28	48	1	130
131	129	129	131	

13	55	54	9	131
36	26	27	40	129
24	43	42	20	129
57	6	7	61	131
130	130	130	130	

4	58	59	8	129
45	23	22	41	131
25	38	39	29	131
56	11	10	52	129
130	130	130	130	

64	17	37	12	130
3	46	31	50	130
2	47	30	51	130
60	21	33	16	130
129	131	131	129	

49	26	39	16	130
5	27	38	60	130
1	42	23	64	130
53	43	22	12	130

KY 180 180 180 180

49	32	44	5	130
14	35	18	63	130
15	34	19	62	130
53	28	48	1	130
131	129	129	131	

9	40	20	61	130
54	27	42	7	130
55	26	43	6	130
13	36	24	57	130
131	129	129	131	

8	41	29	52	130
59	22	39	10	130
58	23	38	11	130
4	45	25	56	130
129	131	131	129	

64	17	37	12	130
3	46	31	50	130
2	47	30	51	130
60	21	33	16	130
129	131	131	129	

49	27	38	16	130
5	42	23	60	130
1	43	22	64	130
53	26	39	12	130

P

LY 180 270 270 180

49	32	44	5	130
14	35	18	63	130
15	34	19	62	130
53	28	48	1	130
131	129	129	131	

61	7	6	57	131
20	42	43	24	129
40	27	26	36	129
9	54	55	13	131
130	130	130	130	

52	10	11	56	129
29	39	38	25	131
41	22	23	45	131
8	59	58	4	129
130	130	130	130	

64	17	37	12	130
3	46	31	50	130
2	47	30	51	130
60	21	33	16	130
129	131	131	129	

49	42	23	16	130
5	43	22	60	130
1	26	39	64	130
53	27	38	12	130

MY 270 0 0 270

53	15	14	49	131
28	34	35	32	129
48	19	18	44	129
1	62	63	5	131
130	130	130	130	

57	24	36	13	130
6	43	26	55	130
7	42	27	54	130
61	20	40	9	130
131	129	129	131	

56	25	45	4	130
11	38	23	58	130
10	39	22	59	130
52	29	41	8	130
129	131	131	129	

60	2	3	64	129
21	47	46	17	131
33	30	31	37	131
16	51	50	12	129
130	130	130	130	

53	43	22	12	130
49	26	39	16	130
5	27	38	60	130
1	42	23	64	130

NY 270 90 90 270

53	15	14	49	131
28	34	35	32	129
48	19	18	44	129
1	62	63	5	131
130	130	130	130	

13	55	54	9	131
36	26	27	40	129
24	43	42	20	129
57	6	7	61	131
130	130	130	130	

4	58	59	8	129
45	23	22	41	131
25	38	39	29	131
56	11	10	52	129
130	130	130	130	

60	2	3	64	129
21	47	46	17	131
33	30	31	37	131
16	51	50	12	129
130	130	130	130	

53	26	39	12	130
49	27	38	16	130
5	42	23	60	130
1	43	22	64	130

P

OY 270 180 180 270

53	15	14	49	131
28	34	35	32	129
48	19	18	44	129
1	62	63	5	131
130	130	130	130	

9	40	20	61	130
54	27	42	7	130
55	26	43	6	130
13	36	24	57	130
131	129	129	131	

8	41	29	52	130
59	22	39	10	130
58	23	38	11	130
4	45	25	56	130
129	131	131	129	

60	2	3	64	129
21	47	46	17	131
33	30	31	37	131
16	51	50	12	129
130	130	130	130	

53	27	38	12	130
49	42	23	16	130
5	43	22	60	130
1	26	39	64	130

PY 270 270 270 270

53	15	14	49	131
28	34	35	32	129
48	19	18	44	129
1	62	63	5	131
130	130	130	130	

61	7	6	57	131
20	42	43	24	129
40	27	26	36	129
9	54	55	13	131
130	130	130	130	

52	10	11	56	129
29	39	38	25	131
41	22	23	45	131
8	59	58	4	129
130	130	130	130	

60	2	3	64	129
21	47	46	17	131
33	30	31	37	131
16	51	50	12	129
130	130	130	130	

53	43	22	12	130
49	42	23	16	130
5	26	39	60	130
1	27	38	64	130

	Rotatable Cube	Z ASPECT	Diagonals

C   AC   AC   C

AZ   0   0   0   0

Rotatable Cube:

1	62	63	5	131
57	6	7	61	131
56	11	10	52	129
16	51	50	12	129
130	130	130	130	

48	19	18	44	129
24	43	42	20	129
25	38	39	29	131
33	30	31	37	131
130	130	130	130	

Z ASPECT:

28	34	35	32	129
36	26	27	40	129
45	23	22	41	131
21	47	46	17	131
130	130	130	130	

53	15	14	49	131
13	55	54	9	131
4	58	59	8	129
60	2	3	64	129
130	130	130	130	

Diagonals:

1	43	22	64	130
5	42	23	60	130
12	39	26	53	130
16	38	27	49	130

BZ   0   90   90   0

Rotatable Cube:

1	62	63	5	131
57	6	7	61	131
56	11	10	52	129
16	51	50	12	129
130	130	130	130	

44	20	29	37	130
18	42	39	31	130
19	43	38	30	130
48	24	25	33	130
129	129	131	131	

Z ASPECT:

32	40	41	17	130
35	27	22	46	130
34	26	23	47	130
28	36	45	21	130
129	129	131	131	

53	15	14	49	131
13	55	54	9	131
4	58	59	8	129
60	2	3	64	129
130	130	130	130	

Diagonals:

1	42	23	64	130
5	39	26	60	130
12	38	27	53	130
16	43	22	49	130

CZ   0   180   180   0

Rotatable Cube:

1	62	63	5	131
57	6	7	61	131
56	11	10	52	129
16	51	50	12	129
130	130	130	130	

37	31	30	33	131
29	39	38	25	131
20	42	43	24	129
44	18	19	48	129
130	130	130	130	

Z ASPECT:

17	46	47	21	131
41	22	23	45	131
40	27	26	36	129
32	35	34	28	129
130	130	130	130	

53	15	14	49	131
13	55	54	9	131
4	58	59	8	129
60	2	3	64	129
130	130	130	130	

Diagonals:

1	39	26	64	130
5	38	27	60	130
12	43	22	53	130
16	42	23	49	130

DZ   0   270   270   0

Rotatable Cube:

1	62	63	5	131
57	6	7	61	131
56	11	10	52	129
16	51	50	12	129
130	130	130	130	

33	25	24	48	130
30	38	43	19	130
31	39	42	18	130
37	29	20	44	130
131	131	129	129	

Z ASPECT:

21	45	36	28	130
47	23	26	34	130
46	22	27	35	130
17	41	40	32	130
131	131	129	129	

53	15	14	49	131
13	55	54	9	131
4	58	59	8	129
60	2	3	64	129
130	130	130	130	

Diagonals:

1	38	27	64	130
5	43	22	60	130
12	42	23	53	130
16	39	26	49	130

EZ   90   0   0   90

Rotatable Cube:

16	56	57	1	130
51	11	6	62	130
50	10	7	63	130
12	52	61	5	130
129	129	131	131	

48	19	18	44	129
24	43	42	20	129
25	38	39	29	131
33	30	31	37	131
130	130	130	130	

Z ASPECT:

28	34	35	32	129
36	26	27	40	129
45	23	22	41	131
21	47	46	17	131
130	130	130	130	

60	4	13	53	130
2	58	55	15	130
3	59	54	14	130
64	8	9	49	130
129	129	131	131	

Diagonals:

16	43	22	49	130
1	42	23	64	130
5	39	26	60	130
12	38	27	53	130

FZ   90   90   90   90

Rotatable Cube:

16	56	57	1	130
51	11	6	62	130
50	10	7	63	130
12	52	61	5	130
129	129	131	131	

44	20	29	37	130
18	42	39	31	130
19	43	38	30	130
48	24	25	33	130
129	129	131	131	

Z ASPECT:

32	40	41	17	130
35	27	22	46	130
34	26	23	47	130
28	36	45	21	130
129	129	131	131	

60	4	13	53	130
2	58	55	15	130
3	59	54	14	130
64	8	9	49	130
129	129	131	131	

Diagonals:

16	42	23	49	130
1	39	26	64	130
5	38	27	60	130
12	43	22	53	130

GZ   90   180   180   90

Rotatable Cube:

16	56	57	1	130
51	11	6	62	130
50	10	7	63	130
12	52	61	5	130
129	129	131	131	

37	31	30	33	131
29	39	38	25	131
20	42	43	24	129
44	18	19	48	129
130	130	130	130	

Z ASPECT:

17	46	47	21	131
41	22	23	45	131
40	27	26	36	129
32	35	34	28	129
130	130	130	130	

60	4	13	53	130
2	58	55	15	130
3	59	54	14	130
64	8	9	49	130
129	129	131	131	

Diagonals:

16	39	26	49	130
1	38	27	64	130
5	43	22	60	130
12	42	23	53	130

HZ   90   270   270   90

Rotatable Cube:

16	56	57	1	130
51	11	6	62	130
50	10	7	63	130
12	52	61	5	130
129	129	131	131	

33	25	24	48	130
30	38	43	19	130
31	39	42	18	130
37	29	20	44	130
131	131	129	129	

Z ASPECT:

21	45	36	28	130
47	23	26	34	130
46	22	27	35	130
17	41	40	32	130
131	131	129	129	

60	4	13	53	130
2	58	55	15	130
3	59	54	14	130
64	8	9	49	130
129	129	131	131	

Diagonals:

16	38	27	49	130
1	43	22	64	130
5	42	23	60	130
12	39	26	53	130

P

IZ 180 0 0 180

12	50	51	16	129
52	10	11	56	129
61	7	6	57	131
5	63	62	1	131
130	130	130	130	

48	19	18	44	129
24	43	42	20	129
25	38	39	29	131
33	30	31	37	131
130	130	130	130	

28	34	35	32	129
36	26	27	40	129
45	23	22	41	131
21	47	46	17	131
130	130	130	130	

64	3	2	60	129
8	59	58	4	129
9	54	55	13	131
49	14	15	53	131
130	130	130	130	

12	43	22	53	130
16	42	23	49	130
1	39	26	64	130
5	38	27	60	130

JZ 180 90 90 180

12	50	51	16	129
52	10	11	56	129
61	7	6	57	131
5	63	62	1	131
130	130	130	130	

44	20	29	37	130
18	42	39	31	130
19	43	38	30	130
48	24	25	33	130
129	129	131	131	

32	40	41	17	130
35	27	22	46	130
34	26	23	47	130
28	36	45	21	130
129	129	131	131	

64	3	2	60	129
8	59	58	4	129
9	54	55	13	131
49	14	15	53	131
130	130	130	130	

12	42	23	53	130
16	39	26	49	130
1	38	27	64	130
5	43	22	60	130

KZ 180 180 180 180

12	50	51	16	129
52	10	11	56	129
61	7	6	57	131
5	63	62	1	131
130	130	130	130	

37	31	30	33	131
29	39	38	25	131
20	42	43	24	129
44	18	19	48	129
130	130	130	130	

17	46	47	21	131
41	22	23	45	131
40	27	26	36	129
32	35	34	28	129
130	130	130	130	

64	3	2	60	129
8	59	58	4	129
9	54	55	13	131
49	14	15	53	131
130	130	130	130	

12	39	26	53	130
16	38	27	49	130
1	43	22	64	130
5	42	23	60	130

P

LZ 180 270 270 180

12	50	51	16	129
52	10	11	56	129
61	7	6	57	131
5	63	62	1	131
130	130	130	130	

33	25	24	48	130
30	38	43	19	130
31	39	42	18	130
37	29	20	44	130
131	131	129	129	

21	45	36	28	130
47	23	26	34	130
46	22	27	35	130
17	41	40	32	130
131	131	129	129	

64	3	2	60	129
8	59	58	4	129
9	54	55	13	131
49	14	15	53	131
130	130	130	130	

12	38	27	53	130
16	43	22	49	130
1	42	23	64	130
5	39	26	60	130

MZ 270 0 0 270

5	61	52	12	130
63	7	10	50	130
62	6	11	51	130
1	57	56	16	130
131	131	129	129	

48	19	18	44	129
24	43	42	20	129
25	38	39	29	131
33	30	31	37	131
130	130	130	130	

28	34	35	32	129
36	26	27	40	129
45	23	22	41	131
21	47	46	17	131
130	130	130	130	

49	9	8	64	130
14	54	59	3	130
15	55	58	2	130
53	13	4	60	130
131	131	129	129	

5	43	22	60	130
12	42	23	53	130
16	39	26	49	130
1	38	27	64	130

NZ 270 90 90 270

5	61	52	12	130
63	7	10	50	130
62	6	11	51	130
1	57	56	16	130
131	131	129	129	

44	20	29	37	130
18	42	39	31	130
19	43	38	30	130
48	24	25	33	130
129	129	131	131	

32	40	41	17	130
35	27	22	46	130
34	26	23	47	130
28	36	45	21	130
129	129	131	131	

49	9	8	64	130
14	54	59	3	130
15	55	58	2	130
53	13	4	60	130
131	131	129	129	

5	42	23	60	130
12	39	26	53	130
16	18	27	49	110
1	43	22	64	130

P

OZ 270 180 180 270

5	61	52	12	130
63	7	10	50	130
62	6	11	51	130
1	57	56	16	130
131	131	129	129	

37	31	30	33	131
29	39	38	25	131
20	42	43	24	129
44	18	19	48	129
130	130	130	130	

17	46	47	21	131
41	22	23	45	131
40	27	26	36	129
32	35	34	28	129
130	130	130	130	

49	9	8	64	130
14	54	59	3	130
15	55	58	2	130
53	13	4	60	130
131	131	129	129	

5	39	26	60	130
12	38	27	53	130
16	43	22	49	130
1	42	23	64	130

PZ 270 270 270 270

5	61	52	12	130
63	7	10	50	130
62	6	11	51	130
1	57	56	16	130
131	131	129	129	

33	25	24	48	130
30	38	43	19	130
31	39	42	18	130
37	29	20	44	130
131	131	129	129	

21	45	36	28	130
47	23	26	34	130
46	22	27	35	130
17	41	40	32	130
131	131	129	129	

49	9	8	64	130
14	54	59	3	130
15	55	58	2	130
53	13	4	60	130
131	131	129	129	

5	38	27	60	130
12	43	22	53	130
16	42	23	49	130
1	39	26	64	130

List of Figures

About the Author

Kenneth Kelsey is one of the dwindling number code-breakers of WWII. He is a retired Chartered Secretary and Barrister and lives in Dorking, Surrey. He has written three books of Number Puzzles brought together in an omnibus volume, 'The Ultimate Book of Number Puzzles', all published by Random House. He has also written a comic novel of life in a fictional English village in the 1950s entitled 'The Nutcombe Papers'; an adventure story for young schoolgirls entitled 'Sophie's Odyssey' and 'A Book of Humorous Poems' all three available through Amazon.